Parallel Computing in Structural Engineering

Ozgur Kurc

Parallel Computing in Structural Engineering

A Substructure Based Approach

VDM Verlag Dr. Müller

Imprint

Bibliographic information by the German National Library: The German National Library lists this publication at the German National Bibliography; detailed bibliographic information is available on the Internet at http://dnb.d-nb.de.

Cover image: www.purestockx.com

Publisher:
VDM Verlag Dr. Müller Aktiengesellschaft & Co. KG , Dudweiler Landstr. 125 a, 66123 Saarbrücken, Germany,
Phone +49 681 9100-698, Fax +49 681 9100-988,
Email: info@vdm-verlag.de

Zugl.: Atlanta USA, Georgia Institute of Technology, Diss., 2005.

Produced in USA and UK by:
Lightning Source Inc., La Vergne, Tennessee, USA
Lightning Source UK Ltd., Milton Keynes, UK
BookSurge LLC, 5341 Dorchester Road, Suite 16, North Charleston, SC 29418, USA

ISBN: 978-3-8364-6088-0

TABLE OF CONTENTS

CHAPTER 1 INTRODUCTION 1

1.1 PROBLEM DEFINITION..1

1.2 RELATED WORK 7

1.2.1 Parallel Environments 7

1.2.1.1 High Performance Computers...7

1.2.1.2 Beowulf Clusters..9

1.2.1.3 Grid Computing 11

1.2.2 Linear Solution of a System of Equations 14

1.2.2.1 Parallel Solution Algorithms – Direct Methods 16

1.2.2.2 Parallel Solution Algorithms – Iterative Methods 26

1.2.3 Domain Partitioning 31

1.2.3.1 Static Partitioning..32

1.2.3.2 Deficiencies of Static Partitioning Algorithms and Improvements 36

1.2.3.3 Dynamic Partitioning 42

1.2.4 Condensation Algorithms..46

1.2.5 Overview 46

1.3 OBJECTIVES AND SCOPE 47

1.4 THESIS OUTLINE 50

CHAPTER 2 INITIAL PARTITIONING AND REPARTITIONING 52

2.1 INTRODUCTION 52

2.2 INITIAL PARTITIONING (STATIC PARTITIONING) 54

2.2.1 METIS Library 58

2.2.1.1 Coarsening Phase...59

2.2.1.2 Initial Partitioning Phase...61

2.2.1.3 Refinement Phase...62

2.3 REPARTITIONING (DYNAMIC PARTITIONING)..63

    2.3.1 Scratch-Remap Algorithms *63*

    2.3.2 Diffusion Algorithms *65*

    2.3.3 PARMETIS Library..*68*

CHAPTER 3 CONDENSATION **73**

    3.1 INTRODUCTION 73

    3.2 METHOD 75

        3.2.1 Theory *75*

        3.2.2 Operation Count *78*

    3.3 CONDENSATION TIME ESTIMATIONS..89

        3.3.1 Time Estimations for Uniform Matrices..*89*

        3.3.2 Time Estimation for Non-uniform Matrices *99*

    3.4 OUT-OF-CORE CONDENSATION 104

    3.5 CONCLUSIONS..106

CHAPTER 4 WORKLOAD BALANCING FOR DIRECT CONDENSATION...........................**108**

    4.1 INTRODUCTION 108

    4.2 METHOD 111

        4.2.1 Workload Balancing Algorithm *111*

        4.2.2 Data Structure..*113*

        4.2.3 Initial Partitioning *115*

        4.2.4 Data Distribution *116*

        4.2.5 Numbering of Substructure Vertices *117*

        4.2.6 Vertex weight definitions..*118*

        4.2.7 Repartitioning *121*

        4.2.8 Convergence Criteria and Choosing Partitioning for Solution *122*

        4.2.9 Interface Element Assignment..*123*

    4.3 ILLUSTRATIVE EXAMPLES AND DISCUSSIONS 125

4.3.1 2D Square Mesh Model..*125*

4.3.1.1 Substructures during Iterations... 125

4.3.1.2 Effect of Interface Element Assignment 133

4.3.2 Three Story Building *134*

4.3.2.1 Substructures Before and After Iterations 134

4.3.2.2 Effect of Interface Element Assignment 140

4.4 EXAMPLE PROBLEMS AND DISCUSSIONS 145

4.4.1 2D Square Mesh..*146*

4.4.2 Half Disk *149*

4.4.3 Bridge Deck...*151*

4.4.4 High-rise Building I & II..*154*

4.4.5 Nuclear Waste Plant *157*

4.5 CONCLUSIONS..158

CHAPTER 5 INTERFACE SOLUTION ALGORITHM..**161**

5.1 INTRODUCTION 161

5.2 PARALLEL VARIABLE BAND SOLVER 163

5.2.1 Serial Implementation *163*

5.2.2 Operation Count for Serial Solution *165*

5.2.3 Parallel Implementation...*168*

5.2.4 Operation Count for Parallel Solution...*171*

5.2.5 Communication Cost..*173*

5.3 PARALLEL SOLUTION TIME ESTIMATIONS 177

5.3.1 Computation Time..*177*

5.3.2 Communication Time *181*

5.3.3 Performance Analysis *191*

5.3.4 Interface Solution of Actual Models...*196*

5.3.5 Performance Comparison *197*

5.4 DISCUSSION OF RESULTS AND CONCLUSIONS 201

CHAPTER 6 IMPLEMENTATION..**204**

6.1 INTRODUCTION 204

6.2 METHOD 210

   6.2.1 Data Preparation *211*

   6.2.2 Parallel Solution Algorithm, Multiple Loading Conditions *215*

   6.2.3 Numbering..*219*

   6.2.4 Parallel Assembly..*223*

6.3 RESULTS AND DISCUSSIONS 225

   6.3.1 2D Square Mesh...*226*

   6.3.2 Half-Disk..*233*

   6.3.3 Bridge Deck...*237*

   6.3.4 High-Rise Building I *240*

   6.3.5 High-Rise Building II..*243*

   6.3.6 Nuclear Waste Plant *245*

6.4 CONCLUSIONS..247

CHAPTER 7 SUMMARY AND RECOMMENDATIONS FOR FUTURE WORK.....................**250**

7.1 SUMMARY..250

7.2 FUTURE WORK 257

APPENDIX A...**261**

A.1 EXAMPLE PROBLEMS 261

   A.1.1 2D Square Mesh *261*

   A.1.2 Half Disk...*262*

   A.1.3 Bridge Deck *263*

   A.1.4 High-rise Building I...*264*

   A.1.5 High-rise Building II..*265*

   A.1.6 Nuclear Waste Plant *266*

APPENDIX B...**267**

B.1 Unified Modeling Language 267

    B.1.1 History *267*

    B.1.2 Structure *268*

        B.1.2.1 Structural Model 269

B.2 Database Structure of the Parallel Solution Program 271

REFERENCES **278**

LIST OF TABLES

TABLE   **PAGE**

3.1   Factorization Times, DEC Cluster   102

3.2   Factorization Times, AFC Cluster   102

3.3   Factorization Times, DELL Cluster   102

3.4   Forward Substitution Times with 50 load cases, AFC Cluster   103

3.5   Forward Substitution Times with 50 load cases, DELL Cluster   104

3.6   Back Substitution Times with 50 load cases, DELL Cluster   104

3.7   Out-of-Core Factorization   105

4.1   Operation Count Values Before and After Interface Elements
       Assignment for Substructures Balanced with Diffusion
       Algorithm   134

4.2   Operation Count Values Before and After Interface Elements
       Assignment for Substructures Balanced with Scratch-Remap
       Algorithm   134

4.3   Condensation Times of 2D Square Mesh on DELL Cluster   147

4.4   Time Spent During Iterations   148

5.1   Properties of the PC Clusters   178

5.2   Computation Speeds for Serial Version   179

5.3   Computation Speeds for Parallel Version   181

TABLE		PAGE
5.4	Broadcast Speeds of DELL Cluster	190
5.5	Interface Factorization Times of 2D Square Mesh	196
5.6	Interface Factorization Times of Nuclear Waste Plant	196
5.7	Computation and Communication Speeds of all Clusters	197
5.8	Detailed Factorization Times of 2D Square Mesh	200
5.9	Detailed Factorization Times of Nuclear Waste Plant	201
6.1	Bandwidth-Number of Interface Equation Ratios with Number of Processors	214
6.2	Data Distribution List	221
6.3	Dof Mapper Lists for Each Processor	222
6.4	Solution with 12-processors, Single Loading Condition	231
6.5	Solution with Initial Substructures, 100 Loading Conditions	231
6.6	Solution with Balanced Substructures, 100 Loading Conditions	232
6.7	Solution with 12-processors, Single Loading Condition	235
6.8	Solution with Initial Substructures, 100 Loading Conditions	236
6.9	Solution with Balanced Substructures, 100 Loading Conditions	236
6.10	Solution with Final Substructures, 100 Loading Conditions	242
6.11	Load Factorization Times for Solution with Multiple Loading Conditions	245

LIST OF FIGURES

FIGURE   **PAGE**

2.1   Characteristics of Various Partitioning Algorithms   56

2.2   An Arbitrary Structure   58

2.3   Graph Representations of an Arbitrary Structure   58

2.4   Multilevel Approach   59

2.5   Matching and Edge Collapsing   60

2.6   Multilevel Diffusion Algorithms   70

3.1   Active Column Storage   76

3.2   An Example Profile of a Stiffness Matrix Assembled for
      Condensation   79

3.3   A Typical Shape of Assembled Substructure Stiffness Matrix   80

3.4   Pseudo Code for Factorization up to ni   81

3.5   Pseudo Code for Factorization for Equations Between 'ni' and
      'n'   85

3.6   Pseudo Code for Forward Substitution   87

3.7   Factorization Times on 166 Mhz DEC Computer   91

3.8   Factorization Times on 400 MHz Celeron Computer (AFC)   93

3.9   Factorization Times on 2.4 GHz Pentium 4 Computer (DELL)   94

FIGURE		PAGE
3.10	Forward and Back Substitution Times on 166 Mhz Computer (DEC)	96
3.11	Forward and Back Substitution Times on 400 Mhz Computer (AFC)	97
3.12	Forward and Back Substitution Times on 2.4 GHz Computer (DELL)	98
3.13	Membrane Problem, 2D 160x160 Mesh with 4 substructures	101
4.1	The Flow Chart of the Workload Balancing Algorithm	112
4.2	The Real Structure and Its Nodal Graph	114
4.3	Nodal Graph and Subgraphs during Iterations	116
4.4	Equation Numbering of Substructures Before and After Modifications	120
4.5	The Actual Structure and Substructures Used during Iterations	124
4.6	Substructures with Interface Elements	124
4.7	Substructures at Each Iteration, Diffusion Algorithm	128
4.8	Substructures at Each Iteration, Scratch-remap Algorithm	132
4.9	The Initial Structure	136
4.10	Initial Partitions Created by METIS Recursive Bisection Partitioning Algorithm	137
4.11	Partitions Balanced with Scratch-Remap Repartitioning Algorithm	139
4.12	First Substructure Before and After Interface Element Assignment	140
4.13	Second Substructure Before and After Interface Element Assignment	141

FIGURE		**PAGE**
4.14	Third Substructure Before and After Interface Element Assignment	141
4.15	Fourth Substructure Before and After Interface Element Assignment	142
4.16	Fifth Substructure Before and After Interface Element Assignment	142
4.17	Sixth Substructure Before and After Interface Element Assignment	143
4.18	Seventh Substructure Before and After Interface Element Assignment	143
4.19	Eighth Substructure Before and After Interface Element Assignment	144
4.20	Condensation Times for 2D Square Mesh	149
4.21	Condensation Times for Half Disk	150
4.22	Condensation Times for Half Disk	151
4.23	Condensation Times for Bridge Deck	152
4.24	Bridge Deck Model	153
4.25	High Rise Building I Model	155
4.26	Condensation Times for High-rise Building I	156
4.27	Condensation Times for High-rise Building II	157
4.28	Condensation Times for Nuclear Waste Plant	158
5.1	Variable Band Storage Scheme	164

FIGURE		PAGE
5.2	Cyclic Data Distribution with 3 processors	168
5.3	Pseudo Code for Factorization	169
5.4	Pseudo Code for Forward Substitution	170
5.5	Pseudo Code for Back Substitution	170
5.6	Blocking Send & Receive Results on DEC Cluster	183
5.7	Blocking Send & Receive Results on AFC Cluster	184
5.8	Blocking Send & Receive Results on DELL Cluster	185
5.9	Broadcast Times on DEC Cluster	187
5.10	Broadcast Times on AFC Cluster	188
5.11	Broadcast Times on DELL Cluster	189
5.12	Factorization Times with DEC Cluster	193
5.13	Factorization Times with AFC Cluster	194
5.14	Factorization Times with DELL Cluster	195
5.15	Speed-up values at DEC, AFC and DELL clusters	199
6.1	Solution Framework Structure	210
6.2	Edges at Substructure Interfaces	213
6.3	Solution Algorithm for Problems with Multiple Loading Conditions	216

FIGURE		PAGE
6.4	Local Numbering of Substructures	219
6.5	Interface Numbering of Substructures	220
6.6	Distributed Numbering of Interface Stiffness Matrix	221
6.7	Pseudo Code for Parallel Assembly Algorithm	223
6.8	Interface Solution with Initial Substructures	227
6.9	Interface Solution with Substructures Created with Diffusion Algorithm	227
6.10	Interface Solution with Substructures Created with Scratch-remap Algorithm	228
6.11	Parallel Solution Times and Speed-up Values for 2D Square Mesh Model	229
6.12	Parallel Solution Times and Speed-up Values for Half Disk Model	234
6.13	Parallel Solution Times and Speed-up Values for Bridge Deck Model	238
6.14	Parallel Solution Times and Speed-up Values for High-Rise Building I Model	241
6.15	Parallel Solution Times with Multiple Loading Conditions for High-Rise Building I Model	243
6.16	Parallel Solution Times and Speed-up Values for High-Rise Building II Model	246
6.17	Parallel Solution Times and Speed-up Values for Nuclear Waste Plant Model	248
A.1	2D Square Mesh Model	260
A.2	Half-disk Model	261
A.3	Bridge Deck Model	262

FIGURE		PAGE
A.4	High-rise Building I Model	263
A.5	High-rise Building II Model	264
A.6	Nuclear Waste Plant Model	265
B.1	The Class Diagram for the Global System	271
B.2	The Class Diagram for the Mechanical Library	272
B.3	The Class Diagram for the Substructure Class	274
B.4	The Substructure Representation of a Structure	275
B.5	The Class Diagram for Parallel Substructure Solver Class	276

CHAPTER 1

INTRODUCTION

1.1 Problem Definition

The rapid developments which have occurred in computer hardware and software technology over the last two decades have made computers an essential and indispensable tool for structural engineering. Today, computers are being extensively utilized in almost every step of the structural design process. Structural engineering software and computer hardware not only speed-up the computations and increase the productivity of the engineer but also improve the quality of design, if used properly. The continued development of computer hardware and improvements in the computer price/performance ratio coupled with improvements in the usability and functionality of structural engineering software have raised the expectations of structural engineers to the point where they appear to believe that any problem and any size model can be analyzed and designed, regardless of the size or complexity of the model. These detailed models could not even be imagined to be solved a couple of decades ago. Moreover, more sophisticated solution techniques are being used and the engineers expect results to be available as soon as possible – within hours instead of overnight. Thus, the race between the needs of the user and technological improvements continues to create a demand for faster computing.

The design of a structure is an iterative process with the structure subjected to many modifications. The reasons of these modifications are not only due to strength, stability, and serviceability requirements but also due to architectural, economical, and manufacturing needs which involve different professions and approaches in the process. For each modification, the structural model may need to be updated, re-analyzed and the structural components redesigned. The result may be a long and time-consuming process.

The analysis/design of a structure can be examined in four basic steps. First is the discretization step where an approximate mathematical model of the structure is created. Then comes the analysis step where the effects of any kind of possible external disturbances on the structure are computed. As the model and the analysis method become more detailed, the results better approximate the actual behavior providing the engineer has made the correct modeling assumptions. Next, the structural components are sized and shaped according to the results of these calculations. These three steps will be repeated many times during the design. Finally, as the design finalizes, the last step, detailing, is performed and the structural drawings and specifications are prepared.

From a computational point of view, the analysis step consumes a considerable amount of computational resources. As the size of the model becomes larger, the finite element procedure requires more time to solve the equations. Depending on the type of problem, various solution techniques are utilized when using the finite elements method. These solution techniques vary depending on the type of analysis being performed: static analysis with many loading conditions, dynamic analysis, and nonlinear static analysis with geometric and material nonlinearities.

Today, the most common analysis method utilized during the design of a structure is static linear analysis where the structure's response to every possible loading condition is computed. The number of loading conditions may be large (over 100); therefore, solving the structure and calculating the element forces for each loading can be very time-consuming.

Similarly, in earthquake engineering, a detailed dynamic analysis has to be performed for most of the structures in highly seismic zones. One of the methods used for that purpose is the linear time history analysis where the response of the structure subjected to an earthquake load is calculated at discrete time intervals. In implicit time history analysis, first, for the ith time increment, the earthquake forces and the dynamic effects are converted into static loads. Then, by constructing the dynamic equilibrium equation for the $i+1^{th}$ increment, the deflections, velocities, and accelerations for that time increment are calculated. This step is simply the linear solution of the system. Hence, the implicit dynamic analysis method may also be considered as the solution of a linear system with repetitive right hand sides.

The solution methods for nonlinear analyses are actually trial and error methods. Although each method has different approach for the solution, all of them first obtain an approximate solution. Then, by using these results, they calculate the amount of error and try to minimize it. One of the popular nonlinear solution approaches is the modified Newton-Raphson method where the same stiffness matrix is used during the iterations until the convergence for the current load increment is obtained. In other words, the same system is solved repetitively with different right hand sides for each load increment.

The analysis steps of all the methods mentioned above consist of three steps. First, the element stiffness matrices are calculated, the loadings are converted to nodal loads, and global stiffness and force matrices are formed. In addition to these, if dynamics effects are considered, mass and possibly damping matrices will be formed. The second step is the solution where the displacements for each load vector are calculated. The third and the final step is the computation of the element forces and stresses.

From a numerical analysis point of view, the solution steps of the above analysis methods converge to the same problem, i.e. the solution of a linear system with multiple or repetitive right hand sides. This step was considered as the most time consuming step and a considerable research [67, 79, 80, 101] has been conducted to speed-up the solution step without any significant loss of accuracy. In addition to that, the time for the stiffness matrix generation and element force and stress computations may consume significant amount of time. As a result, a complete analysis approach that is able to speed-up all of the analysis steps of large structural models in an accurate and robust way is still an important need in the civil engineering industry. Such an improvement would speed-up the design process that will enable the structural engineer to evaluate more alternative designs and help alleviate project deadlines.

Today, computers are much faster and more affordable. For example, in 1988, the cost of a 0.25 MHz personal computer was around $3,000 and the cost of a 3MHz workstation was near $20,000 [109]. In the following years, there were great developments both in supercomputer and microprocessor technology. Similar developments were also observed

4

in the memory and disk space areas. In the late 80's engineers used PC's with memory sizes in the order of Kbytes and storage space sizes of Mbytes. In 2005, a personal computer often has a 1 Gbyte of memory and at least 200 Gbytes of hard drive for a price often less than $500. Home computers often have 3 GHz processors. Furthermore even small design offices have a computer network system.

Likewise, high performance computer systems have gone through many evolutionary changes. Parallel computer systems are not new; they existed even before the Iliac IV system [104] of 1968. Since then, together with the technological advancements in the computer hardware area, there have been a lot of changes in parallel computer architecture. Unfortunately, a preferred architecture for parallel systems has not emerged. The difficulty in designing parallel system architecture not only arises from the hardware design but also from the design and development of the operating system and compilers. These concepts are inseparable. That is the main reason why research [106] is still being conducted on various parallel hardware alternatives.

Currently, parallel computers are also more available and affordable. Many home computer systems have a socket for a second CPU. Similarly, the operating systems are capable of using more than one processor. Moreover, in most of the businesses and universities, nearly all of the computers are connected to a network. Supercomputers push the boundaries of imaginable speed and capacity even further. The fastest computer system in the world in 2004 is located in Rochester, USA [119]; it has 32,768 nodes of 700 MHz processors. The fastest network of computers is currently located at the Barcelona Super Computer Center and consists of 4,536 64-bit 2.2GHz PPC970FX processors [119].

Given this variety of computer architectures and the rate of development in computer technology, it becomes very difficult to decide on the optimum parallel computer system for a specific purpose. A number variables affect this decision: computing power, storage requirements, cost including the maintenance and software prices, ease of use, upgrading options, and amount of utilization, etc. Software developers face similar problems. The parallel properties of the computer system, such as the communication topology, memory access, and communication speed affect the design of a parallel algorithm. As a result, the first step of parallel programming should be a decision on the targeted computer architectures.

The most readily available computer system in civil engineering design offices for parallel computing is the network of PC's with the Windows operating system. A parallel solution system that is optimized for such computer networks will benefit the structural engineers considerably. Not only the time spent during the analysis will decrease but also, their existing computer system will be utilized more efficiently without the need of purchasing additional hardware.

As a result, the civil structural engineering industry will benefit significantly from a solution algorithm that is capable of utilizing the existing computer systems at many design offices, optimized for linear static solution with multiple loading conditions, and able to decrease the time spent during the analysis considerably. That is why this study will focus on parallel linear solution techniques on PC clusters in order to provide an economical, functional, and effective tool to structural engineers.

1.2 Related Work

1.2.1 Parallel Environments

1.2.1.1 High Performance Computers

The high performance computers have variety of hardware architectures that can be classified according to their way of manipulating the instructions and data streams by using the taxonomy proposed by Flynn [110]. The most important high performance computer categories are as follows:

- **SIMD machines:** Single instruction machines that manipulate many data items in parallel. Such machines have large number of processors, ranging from 1,024 to 16,384. Vector processors are one type of SIMD machines.

- **MIMD machines:** These machines execute several instruction streams in parallel on different data. There are many kinds of MIMD systems that can be further classified according to their memory taxonomy as shared and distributed memory machines.

Commercially available machines such as the SGI Origin 3000 and NEC TX7 are examples of the shared memory MIMD machines. In such systems, each CPU accesses the same memory via a type of bus or crossbar. The bus can be thought of a bunch of cables that connects the processors to the memory and peripheral controllers. In these systems, all the communication must be done by using the bus and there can be only be one communication along the bus which is the primary source of bottlenecks. Crossbars, on the other hand, can be visualized as several buses running side by side, connected to all of the processors, peripherals and memory. They allow more than one communication at the same time; however, they may increase the price of the machine significantly. For example, SGI Origin 3000 utilizes an Ω-network type of crossbars where for 'n' processors there are 'log n' switching stages where many data items must compete for any path.

The distributed memory MIMD machines are the fastest growing category of the high performance computers [127] although their programming is much more complex than the shared memory MIMD machines. The distributed memory MIMD machines can have many processors with each processor having its own memory. The processors communicate with each other using messages. The connection between processors can be viewed as a network system. Actually, there are many different interconnection topologies some of which are: bus architecture, crossbar, pipelined multistep communications, multistage interconnection network, mesh and toroid connections, hypercubes and irregular topologies. Each of these approaches has their strengths and weaknesses in areas such as price, scalability, speed, bandwidth, latency, ease of manufacture, etc. When compared with the shared memory machines, the communication speed between their processors is much slower.

The network of computers or the clusters are actually the most basic types of distributed memory systems. All computers are simply connected to the network and all the data transferred across the network by routers, hubs or switches. This is the cheapest way to obtain a high-performance computer system and they can be upgraded or extended for a very low price. Hence, these systems are becoming more popular. On the other hand, it is difficult to obtain high performance due to the relatively slow communication among the computers.

There are also Symmetric Multi-Processing (SMP) machines that could be considered as hybrid systems. In such systems, first the processors in a node are connected with a crossbar, and then the nodes are connected with a less costly network.

The IBM Blue Gene/L [112] is a distributed memory MIMD machine which is the fastest high performance computer in the world in year 2004. Its basic processor speed is 700 MHz. Two of these processors reside on a chip with 4MB L3 cache and 2KB L2 cache memories. These two chips fit on a computer card with 512 MB memory. 16 of these cards fit on a node board, and 32 boards go into a cabinet. Thus a single cabinet contains 2048 CPUs. The system has both 3-D torus and tree network connection topologies. The torus network is utilized for general communications and the tree network is for the collective communications such as broadcasting and reduction operations. The system with 32768 CPUs has attained a speed of 70.71 Tflops on the HPC Linpack benchmark [127].

The NEC Tx7 [118] is a shared memory SMP system. The i9510 model has 32 1.5 GHz Itanium 2 processors. The processors are connected by a flat cross-bar. Its 16 processor model has attained of a speed of 14.5 Gflops for the dense matrix-vector multiplication [127].

1.2.1.2 Beowulf Clusters

According to Brown [105], the accepted definition of the Beowulf cluster is a cluster of computers interconnected with a network having the following characteristics:

- The computers and the network are dedicated to the Beowulf and utilized for high-performance computing.

9

- Both computer and network components are relatively inexpensive, mass-produced and readily available.

- All computers run open-source software.

These requirements focus on the major philosophical motivation of constructing a 'Beowulf cluster' which is to save money. For this reason, a Beowulf cluster is usually constructed of PCs which are connected with standard networks to perform parallel computations.

There are also other characteristics that are still argued to be the additional necessary conditions for Beowulf such as:

- All computers are identical, i.e. configured with the same CPU, motherboard, network, memory etc.

- All computers perform a single computation at a time.

These two additional requirements have a practical basis. If the computer architectures are not identical, the computation time required to complete a given parallel step of the calculation should be predicted very accurately. Moreover, the workload should be distributed to each computer such that each computer completes computational steps approximately at the same time.

The Stone Souper Computer [128] was an example of the Beowulf cluster which was built by using the old computers donated by the individuals from Oak Ridge National Laboratory with no cost. The cluster consists of 133 computers with 80 Intel 486DX-2/66 and 53 Intel Pentium processors, most having 32 MB RAM. The computers were connected with 10Mbit network cards. The cluster was utilized for large-scale landscape analysis and global terrestrial ecosystem carbon model.

The ASCI cluster at Caltech [103] is another Beowulf cluster that is dedicated to small scale problems, long production runs, and code development and validation. The cluster is composed of 100 machines each having an Intel Pentium III 1GHz, 1GB ram, and 30GB hard drives. The computers are connected with Gigabit ethernet and each computer has 100 Mbit network cards.

1.2.1.3 Grid Computing

The idea of grid computing is not new, but its definition has been evolving and extended since the early 1980's [125]. In the early 1980's, the research work on grid computing focused primarily on networked operating systems [111]. In the early nineties, researchers investigated the distributed operating systems [108] and heterogeneous computing [124]. Then, the grid computing was considered analogous to parallel distributed computing where parallel codes ran on distributed resources.

Today's idea of grid computing has several differences from the older distributed paradigm [125]. First of all, the resources of a grid should not subject to a centralized control. In other words, grids integrate the resources belonging to different domains, such as the computer clusters of different companies or different departments in the same organization including universities. Thus, each local cluster has its own account, security and user policies etc. Secondly, grids involve heterogeneity. Instead of having software and hardware homogeneity at each domain, grids attempt to define standard interfaces and protocols for resource sharing, communication etc. Moreover, grids not only involve computers and networks but also specialized scientific equipments. As a final characteristic, the main focus of grid systems should be the user. The grids should be able to use its resources in a coordinated and secure fashion to deliver the service in a certain amount of time.

These differences have the capability of making grids more useful than its predecessors. On the other hand, they create many problems that must be solved before the grids can be utilized other than in academic environments. First of all, grids have to become more functional. Today, they are not utilized to solve a single task using the domains at geographically distant sites. Instead, they are utilized to solve simple parallel application (SETI@home [126]) or for easier resource selection (PACI Genie Work [123]). Moreover, there are not many grid users due to insufficient software tools, security issues, and a lack of standard interfaces. The other primary problem of grids is the unpredictable behavior of resources. For example, the user can not know whether the shared resource is utilized by someone else. Similarly, the bandwidth of the local network varies from time to time. As a result, the grid approach has many challenges. In order to become a functional environment, such problems must be solved.

The Globus project [121], brings several organizations together to develop fundamental technologies to build computational grids. They conduct research in areas such as resource management, security, information services, and data management. They have been developing Globus Toolkit that included software services and libraries for resource monitoring, discovery, and management plus security and file management. They also assist in the planning and building of grid enabled applications.

The NEESgrid [117] links earthquake researchers across the US with computing resources and research equipments. At a case study with NEESgrid [117] software, a

Multi-site Online Simulation Testbed (MOST) [124] experiment was conducted. This experiment linked a physical experiment conducted in the Newmark Civil Engineering Laboratory at the University of Illinois at Urbana-Champaign (UIUC) and at the Structures and Materials Testing Laboratory at the University of Colorado, Boulder (CU) with a numerical simulation software at the National Center for Supercomputing Applications (NCSA), also in Urbana-Champaign. In order to test the structure, Multi-site Pseudo Dynamic Substructure method [116] was applied. In the Multi-site Pseudo Dynamic Substructure method, the test structure was divided into several substructures and each substructure was physically tested or numerically simulated at the same time at different locations in a coordinated manner. In this experiment a two-bay single story frame was divided into two cantilever columns at edges and the central frame. Two cantilever columns tested at two different laboratories while the central frame section was modeled by the simulation software. The tests and simulations were performed simultaneously and one of the NEESgrid [117] tools, NTCP, was utilized for the communication between the actuators and the simulation programs. During the experiment, the structural response was streamed to remote users and simultaneously stored in the main data repository for archiving. The NEESgrid [126] and MOST [124] experiment showed how a grid application could utilize computers and scientific equipments simultaneously and share the real-time response with remote users.

1.2.2 Linear Solution of a System of Equations

A system of linear equations is represented by:

$$Ax = b \tag{1.1}$$

where A is an N × N nonsingular coefficient matrix, x is an N × 1 unknowns vector and b is an N × 1 known right-hand side vector. When there are multiple right-hand sides, the unknowns are computed for each right-hand side vector one-by-one. According to the solution method applied, the type of the coefficient matrix A, may vary as follows:

- Dense or sparse

- Symmetric or unsymmetric

- Positive definite or non-singular

There are different solution methods which work more efficiently depending on the nature of the coefficient matrix A. These methods can be classified into the following two groups although there are methods that utilize the features of both methods:

- **Direct methods**: These methods give the exact solution of a linear system with known number of operations. There are mainly two different approaches in direct solution methods: (1) finding the inverse of the coefficient matrix and multiplying it with the right hand side vector or (2) transforming the coefficient matrix into triangular or diagonal form in order to decrease the coupling between the equations. The first method is seldom used due to the large number of operations. The most commonly used transformation based direct methods are Gauss elimination, LU decomposition, Cholesky decomposition, Gauss-Jordan methods, Givens-rotation based methods, and Householder reductions.

14

- **Iterative methods**: These are trial and error methods. They start with an initial guess of the solution and then they attempt to converge to the correct solution iteratively by refining the solution at each iteration. There are many different methods of iterative solution but the most common ones are: Jacobi method, Gauss-Seidel method, successive over relaxation (SOR) and the conjugate gradient method. Each of these methods have various preconditioning techniques in order to decrease the number of iterations while approaching the real solution.

The stiffness matrices arising from the linear static finite element method are symmetric and positive definite. Moreover, the stiffness matrices are often sparse where most of their elements are zeroes and generally are stored in active column or compressed row format. In the active column storage scheme only the elements up to the first non-zero element in each column are stored in a vector form. The location of the diagonal elements is stored in another vector. In the compressed row format, on the other hand, only the non-zero elements of the matrix are stored. The non-zero elements of the stiffness matrix and their column ids are stored in two different vectors. There is another vector that stores the starting index of each row. The linear solution algorithms differ according to the way the stiffness matrix is stored. In the literature, there has been an extensive research using both active column and compressed row storage schemes [67, 70, 92] on the linear solution of a symmetric-positive definite sparse system.

In the past, the main focus of researchers working on linear solution techniques was to develop serial algorithms that decreased the solution time without affecting the accuracy of the results. The most important issues were the minimization of the number of operations and the development of efficient storage methods to overcome memory

15

limitations. Today, much of the research effort [62, 69, 76, 79, 82, 88, 92] in the linear solution area is being spent on parallel computation as parallel computers have become more available and affordable. Moreover, parallel solution techniques have the capability of decreasing the linear solution time significantly. Factors such as scalability, workload balancing, efficiency, and communication cost are the additional important issues that affect the efficiency of a parallel solution method. For example, the communication cost is the time required to transfer data from one processor to another. Scalability refers to the capability of an algorithm to improve the solution time as the number of processors increases. Thus, a scalable solution method can utilize more processors synchronously and perform the solution much faster. Similarly, workload balancing refers to the equal distribution of computational loads among the processors. Any imbalance in the computational load causes some of the processors to remain idle during the solution which will slow down the solution. If the processors remain less idle, the parallel algorithm will be more efficient. If the parallel algorithm is more efficient, larger speed-up values will be obtained.

Past research has not only investigated the rewriting of the solution algorithms designed for sequential computing but has also developed entirely new strategies and algorithms to take advantage of the characteristics of various parallel computing architectures. Today, there are numerous parallel solution methods that will be examined in two groups: direct and iterative.

1.2.2.1 Parallel Solution Algorithms – Direct Methods

Different versions of direct solution methods have been used extensively in many finite element applications. Direct methods are accurate and robust; however, they require large storage spaces and are not as scalable as iterative methods. There has been considerable research [55, 67, 78, 79] in the literature that has tried to improve the scalability of such methods. The direct solution methods can be initially classified as active-column, sparse and multi-frontal solvers. In addition, there are domain decomposition based solution approaches that can utilize different direct solvers for the local and interface portions of the solution.

Active Column Solvers

Farhat and Wilson [79] presented a parallel active column solver based on LU decomposition method for sparse and dense symmetric systems of linear equations. They designed versions for both distributed and shared memory systems. In the distributed version, they used cyclic data distribution (processor 1 stores columns 1, 1+p, 1+2p, processor 2 stores columns 2, 2+p, 2+2p and so on where p is the total number of processors). The column by column distribution scheme was utilized during factorization and forward substitution. However, during back-substitution, all the coefficients that would be used to calculate the kth displacement were stored in a single processor. Hence, in order to perform back substitution in parallel, these coefficients had to be redistributed. For that reason, they recommended the back-substitution should be performed sequentially on local memory machines. They tested the solution algorithm on two different parallel machines; iPSC, a distributed memory computer with hypercube topology, and Multimax, a shared memory computer. They solved various size matrices and for a dense matrix having 1,200 equations, they obtained 78% and 67% overall efficiency using iPSC with 32 processors and using Multimax with 6 processors, respectively. The efficiency was lower for forward substitution, around 34% for both

17

computers. The results showed that the efficiency of the algorithm improved as the size of the matrices increased.

In another study performed by Farhat [78], the parallel active column solver was redesigned in order to exploit both parallel and vector capabilities of high performance architectures. In this version of the algorithm, first a square block, e.g. rows and columns ranging from 1 to d, of a matrix A was factorized in column-wise fashion (ijk order). Then, the unfactorized columns of the rows from 1 to d, were updated in row-wise fashion (jki order). As a final step, the remaining equations of matrix A were reduced block by block (kji order) with the coefficients of the initial block (rows from 1 to d). This way a loop unrolling of depth d was obtained which improved the parallel efficiency of the algorithm. The speed of the new algorithm was then compared with the other parallel versions of active column solvers and the variable band solver. When the stiffness matrix was dense, the variable band solver outperformed all the other versions of active column solvers. However, as the sparsity of the stiffness matrix increased, Farhat's new algorithm became 2 to 4 times faster than the other solvers using 4 processors.

Sparse Solvers

The solution of sparse systems has been studied extensively with various rearrangements of Gaussian elimination for various parallel architectures. In the review paper written by Duff and van der Vorst [67], the sparse matrix factorization schemes were characterized in three groups: Fan-out [84], fan-in [57] and multifrontal [87]. In the 'fan-out' type of factorization techniques, the data from a single row or a column is used to modify the subsequent rows or columns. For example, in the column-wise fan-out LU factorization technique, first the lower triangular coefficients of a single column are computed, and then the remaining columns are updated by using these coefficients. On the other hand, in the 'fan-in' type of factorization technique, the coefficients from the previous rows and columns are used to update the current row or column.

A fan-out algorithm was developed by George et al. [84] for distributed memory machines. They proposed a subtree-to-subcube mapping to solve large sparse systems with Cholesky decomposition on hypercube machines. This way, the data locality was preserved by assigning subtrees of the elimination tree to contiguous subsets of neighboring processors.

Ashcraft et al. [57] proposed a demand driven approach for sparse systems based on fan-in approach. In their approach, the updates of a given column j are not computed until needed to complete that column, and they are computed by the sending processor instead of the receiving one. As a result, all of a given processor's contributions to the updating of the column in question were combined into a single aggregate update column, which is then transmitted in a single message to the processor containing the target column. This approach effectively decreased the communication frequency and volume.

In another study by Ashcraft [58], a 'fan-both' method was implemented that combined the 'fan-in' and 'fan-out' approaches in a single algorithm. He calculated the communication volumes of all the 'fan family' algorithms and also tested them on several example problems. Within the view of the performed test and analytical results, he noted the potential of the 'fan-both' algorithm to outperform the 'fan-in' and 'fan-out' methods when a large (128-256) number of processors were available.

A full sparse solver was developed by Heath and Raghavan [87] where the framework consisted of four basic steps: ordering, symbolic factorization, numeric factorization, and triangular solution. They used a geometric ordering approach, called Cartesian nested dissection, for reordering the equations that was based on coordinate information for the underlying graph of the matrix. They applied this method until they obtained as many subgraphs as the number of processors. Next, they used symbolic factorization to determine the data fill and the amount of storage space by using a separator tree. Then they utilized multifrontal sparse Cholesky method [88] for numerical factorization. In this method, the solution starts at the leaves of the separator tree, combining portions of the corresponding dense submatrices and propagating the resulting update information upward to a higher level in the tree. As a final step, triangular factorization was performed in the same way as the factorization step. The proposed solver was able to decrease the solution time using 2 to 32 processors on the iPSC/860 computer. After that point, there was not any gain in the solution time as the number of processors increased.

Schreiber [99] performed an analytical study which showed the scalability of the column oriented approach for the solution of sparse systems on distributed memory machines. Schreiber commented that the column-wise approach could be used for moderately parallel machines; however, for massively parallel machines either 2D mapping of dense frontal matrix or a 'fan-out' submatrix Cholesky algorithm should be utilized.

Multifrontal Methods

In frontal methods, originally developed by Irons [90], the factorization proceeds as a sequence of partial factorizations on a full submatrix, called the frontal matrix, of the overall system. Frontal matrices are dense matrices and they can be factorized by using optimized versions of linear algebra subroutines. The multifrontal version of the frontal

20

method is very applicable to parallel computing since it allows processing multiple independent fronts simultaneously on different processors.

The distributed memory machine version of the multifrontal method was implemented by Conroy et al. [65] with an efficient data structure. The MUMPS code [55], that is a part of EU PARASOL Project [96], is another implementation of a parallel multifrontal technique for distributed memory computers. The EU Parasol Project [96] aims to build and test a portable library for solving large sparse systems of equations on distributed memory systems. The MUMPS code [55] consists of three phases. The first one is the analysis of the systems matrix in order to compute the multifrontal tree that will be used during synchronization. Then, the factorization starts. The final step is the solution step where the right hand sides are solved one by one using the factors computed in the factorization step. After testing the code on several test problems, the authors concluded that the current version of the MUMPS [55] code produced comparable speed-up values to the shared memory variants for small number of processors. The performance of the algorithm improved up to 16 processors; after that, no significant speed-up was obtained because of the increased communication cost during stiffness matrix distribution and factorization.

There are two main problems in multifrontal methods; they may require large memory space for in-core storage which is not always possible. Secondly, their parallel efficiency depends on how the elimination trees are constructed. The study performed by Guermouche et al. [83] focused on these problems. They compared the effect of five different reordering algorithms, AMD[56], AMF [55], PORD [98], METIS [30], and SCOTCH [46], on the shape of the corresponding elimination trees, hence, their effects

21

on the memory usage. The METIS [30] and SCOTCH [46] libraries gave wide well-balanced trees where the others gave very deep unbalanced trees with a large number of nodes. In terms of memory usage, deep unbalanced trees were found to be better than the wide ones. They concluded that, however, for parallel cases, the computational scheduling had to be considered since it also had a significant effect on the memory requirement.

Domain Decomposition Approach

Both sparse and multifrontal methods are designed to perform the solution of systems of equations as efficiently as possible. They do not focus on the computations before or after the computations. However, in finite element analysis, stiffness, force matrix generation and computation of element results may take considerable time. For such cases, substructuring methods are very effective.

Substructuring methods can be applied to the multifrontal technique. In such a case, instead of creating elimination trees, the underlying domain is partitioned into subdomains and frontal decompositions are performed on each domain separately. Duff and Scott [69] presented an example of this approach. They solved grid problems on CRAY and DEC Alpha machines, and obtained super-linear speed-up values as the number of subdomains increased. Having larger speed-up values larger than the number of processors was due to the reduction in the number of floating point operations due to the decrease in the size of subdomains.

The classical method of substructuring was implemented using a parallel finite element solution by Farhat et al. [81]. Their method began by partitioning the structure

22

into subdomains. For each subdomain, the stiffness matrix and force vectors were formed by first numbering the internal degrees of freedom and then the interface degrees of freedom. The internal equations were transferred to the interfaces by static condensation. For the solution of the interface problem, the row oriented formulation of the LDL^T method was utilized. Two different data storage schemes were used, cyclic row-wise and cyclic block. The algorithm was tested on a hypercube iPSC parallel computer and the algorithm ran with 90% efficiency up to 6 processors. The efficiency dropped to 70% as the number of processors increased. However, when a larger problem was solved, the efficiency ranged between 80%-90%. The paper by Farhat and Wilson [80], described the architecture of the above computer program. The proposed structure had a more generalized format which would be used for dynamic and non-linear problems also.

In another study done by Baugh and Sharma [60], the domain decomposition method was used for solving linear equations on a network of workstation systems. They compared five different algorithms based on direct, iterative and hybrid methods. In the direct approach, the partitions were first condensed with a direct static condensation method and a direct solution was performed at the interface. In the iterative approach, they solved the system globally by using two different versions of the conjugate gradient method. The hybrid approach used direct condensation and parallel and serial versions of conjugate gradient method for the interface problem. They then solved a rectangular membrane problem. The test results showed that the iterative solution methods were outperformed by the direct solution methods for this specific problem.

Fulton and Su [82] implemented the substructuring method on a shared memory parallel computer. They used the active column storage scheme to store the substructure level stiffness matrix. During the condensation, the internal equations were first numbered and then the interface equations. The interface stiffness matrix was kept in the shared memory and the contribution of each substructure was assembled to the interface stiffness matrix according to the correct location determined during the renumbering phase. In order to balance the various computational loads for the condensation phase of each substructure, more processors were assigned to the substructures which were estimated to require more computation. The proposed approach performed much better than the parallel global solution algorithm.

The following research by Chuang and Fulton [64] compared the active column and sparse matrix storage schemes using the Cholesky method of condensation. Moreover, they tested the different matrix reordering methods on the execution times. The results showed that the performance of both storage schemes depended on the problem under consideration but their study demonstrated the potential of the sparse matrix algorithms.

The other paper by Synn and Fulton [100] searched for answers to the following issues: direct versus iterative solution, the optimum number of processors for the parallel matrix decomposition, workload balancing, and which solution type for a particular problem. They recommended the direct solution method due to the fewer number of operations even though iterative methods were scalable. The load balancing during condensation step was provided by assigning more processors to the subdomains estimated to have larger number of equations and higher bandwidths. Moreover, they derived operation count equations to help the user to estimate the optimum number of processors and to choose whether to use the global solution instead of a substructure solution.

An object-oriented database structure was proposed by Hsieh et al. [89] that could be used in parallel finite element codes for structural engineering applications. They preferred substructuring approach with direct solvers in their code which utilized the parallel matrix library developed by Modak et al. [93]. In this library, the linear solution algorithm was based on active column matrices and used the Cholesky decomposition method. They tested the interface solution algorithm on Sun Sparc 10 and Intel Paragon machines. Small speed-ups were obtained for the factorization phase during the interface solution. Moreover, the forward and back substitution times remained constant as the number of processors increased.

Yang and Hsieh [102] focused on the load balancing problems for direct substructuring methods and presented an iterative partition optimization method. The proposed method starts with partitioning the domain by using METIS [30] graph partitioning library. After the equations at each partition are reordered, the approximate condensation times were computed. The workload imbalance was smoothed by either changing the nodal weights and repartitioning it (IMP-MNM) or by migrating element among the partitions (IMP-MMJ). They used METIS [30] for the IMP-MNM method, JOSTLE [52] for the IMP-MMJ and a sparse solution approach for condensation. They performed a linear static analysis on a four node PC cluster. The results showed that IMP-MMJ method balanced the condensation times much better than the IMP-MNM method. However, the IMP-MMJ method consumed significantly more time than the IMP-MNM method. Hence, the IMP-MMJ method was considered to be more suitable for dynamics or non-linear problems.

Escaig et al. [71] presented a multilevel domain decomposition method with a direct solver for the interface problem. They first partitioned the structure in such a way that the number of subdomains was larger than the number of processors. During the parallel solution, the subdomains were condensed by the first available processor. This way it was possible to balance the workload among the processors. However, as the number of subdomains increased, the size of the interface problem also increased. They tested their algorithm both in shared and distributed memory architectures. Although they obtained good results for shared memory architectures, the performance dropped as the number of processors increased for distributed architectures.

An analytical study performed by Nikishkov et al. [72] examined the parallel performance of the domain decomposition method with LDU based condensation and solution algorithms. They first calculated the number of operations and the communication volumes and estimated the solution times of each algorithm for a square domain problem. Then, they compared the time estimations with the actual values. The predicted values mostly agreed with the actual ones (<5%) and good parallel efficiency 95% with 6 processors, 85% with 8 processors, was obtained.

1.2.2.2 Parallel Solution Algorithms – Iterative Methods

Iterative methods offer a lot of advantages for the solution of large systems. They are highly scalable and they require less memory then the direct solvers. However, in order to decrease the number of iterations, they require a good preconditioner. Moreover, robustness is a concern especially for the domain decomposition approach. Current research [73, 75, 80, 85] has mainly focused on the scalability, preconditioning and improved robustness issues.

The comparative study carried out by Bitzarakis et al. [62] investigated three domain decomposition formulations with preconditioned conjugate gradient method (PCG). The first approach, called global-subdomain implementation (GSI), is a subdomain by subdomain PCG algorithm that is implemented on the global stiffness. In the second approach, called primal subdomain implementation (PSI), the PCG algorithm is applied on the interface problem after eliminating the internal degrees of freedom by using static condensation. The final method, called dual subdomain implementation (DSI), is actually the FETI method introduced by Farhat and Roux [76] which is discussed later in this section. They compared the results of these three approaches on an eight processor machine. DSI algorithm performed the fastest where GSI was the slowest according to the results of the example problems.

Gullerud and Dodds [85] presented a design of a linear preconditioned conjugate gradient solver (LPCG) using an element by element framework and described the implementation within a nonlinear, implicit finite element code. Their method used two levels of mesh decomposition. As a first step, the mesh was partitioned into subdomains to achieve coarse grain parallel execution. Then, each partition was divided into blocks for efficient fine grain parallel computations. Moreover, they presented a new parallel implementation of Hughes-Winget preconditioner for the LPCG. In order to demonstrate the performance of the parallel algorithm, they solved three example problems and compared the results with the direct sparse solver solution. For 32, 48 and 50 processor solutions, the LPCG solver outperformed the direct solver. Moreover, the algorithm demonstrated good parallel performance and scalability.

Farhat and Roux [76] derived a new version of the domain decomposition based solution algorithm for the solution of large systems. The new method, called finite element tearing and interchanging (FETI) method, removes the continuity constraint between the subdomains by introducing a Lagrange multiplier. The solution starts with an arbitrary mesh having floating subdomains. First, the rigid body motion modes are eliminated from each subdomain. Then, the local problem is solved with a direct solution method. Next, the contributions of these rigid body modes are related to the Lagrange multipliers through an orthogonality condition. As a final step, a parallel conjugate gradient method is utilized to solve the coupled system of local rigid modes and Lagrange multipliers. The proposed method requires less communication than the classical substructuring methods because the subdomains which interconnect along one-edge in three dimensional problems and the ones which interconnect along one vertex do not require any interprocessor communication. They compared the results of the FETI method with direct sparse method solutions, and the FETI method was faster than the direct solution for most of the cases.

In the following research, Farhat et al. [75] discussed the numerical and parallel scalability of the FETI method. Parallel scalability characterizes the ability of an algorithm to deliver larger speed-up values for a larger number of processors. On the other hand, the method is considered numerically scalable if the arithmetic complexity grows almost linearly with the problem size. They demonstrated that the FETI method can compute faster when the number of processors was increased for a fixed size problem by performing static solution of a 3D problem. Moreover, as the number of processors increased, the method solved larger problems at a constant CPU time.

M. Bhardway et al. [61], tested the scalability of FETI method on ASCI Option Red supercomputer configured to use 1000 processors. They used FETI as the equation solver for Salinas, a massively parallel implicit structural dynamics code. They first solved two 3D homogeneous benchmark problems. The first structure had a cubic shape and partitioned into $n_x n_x n$ subdomains. The second structure had a rectangular parallelepiped shape and partitioned into $2_x 2_x n$ subdomains. Both problems achieved numerical scalability with a speed up in the range of 700-900 with 1000 processors.

Different versions of the FETI method were proposed by different researchers in order to improve the convergence and robustness properties. Park et al. [97] proposed a different version of FETI algorithm with the following three attributes: an explicit generation of the orthogonal null-space matrix, floating rigid-body modes, and the identification of redundant interface force constraint operator. Farhat et al. [80] modified the previous version of FETI algorithm to provide better numerical scalability for the fourth-order elasticity problems such as shells and plates. In this version, an optional constraint is enforced at each iteration. This way, the new approach converges faster than the previous one.

In another paper by Farhat et al. [73], they presented the second generation of FETI family algorithms which offered more efficiency on a larger number of subdomains, greater robustness, better performance, and more flexibility. They first discussed the robustness and performance issues about the first generation FETI solvers and classified the critical components of a FETI solver as follows: subdomain zero energy modes, coarse problem solver, influence of number of subdomains, and local solvers. The subdomain energy modes have to be predicted correctly; otherwise, the problem can not

29

be solved. In order to overcome this deficiency, they proposed a two step geometric-algebraic method to eliminate and predict any possible zero energy modes. Moreover, instead of using iterative local solvers, Farhat et al. [73] commented about the potential of direct methods, either skyline or sparse, on improving the robustness and performance of FETI solvers although iterative solvers were very attractive since they reduced the storage requirement. For problems with multiple or repetitive right hand sides, the standard CG algorithm was not suitable due to the fact that there was not an efficient preconditioning technique for a large number of subdomains. Hence, for such cases, the coarse system matrices are also factored in parallel using direct methods. However, this time the forward and back substitutions were performed sequentially that limited the scalability of the current generation FETI solvers. The second generation FETI solvers were tested on various problems and CPU times were compared with the parallel direct sparse solvers. One level FETI solvers outperformed the sparse solvers but the sparse solver outperformed the two level FETI solution for a low number of processors.

One of the biggest challenges for an iterative solution approach is the problems having multiple or repetitive right hand side cases (multiple load vectors). The direct solvers are well-suited for such cases since once the system matrix is factorized, the solution is obtained with relatively inexpensive forward and backward substitutions. In such cases, the iterative solvers often start from scratch for each loading case. Farhat et al. [76] presented a method to overcome this difficulty for the FETI method. They formulated the overall problem as a minimization problem over K-orthogonal and supplementary subspaces. The resulting method was scalable in the fine granularity regime. The authors demonstrated its performance on dynamic transient and static multiple loading condition analyses. The results were improved over the direct forward and backward substitution solutions because as the number of solved right hand side increased, the number of iterations decreased. In another test performed by Bitzarakis et al. [62] a static linear problem was solved with ten load cases in which each load case were selected to excite a different dynamic mode of a structure. In this case, the improvements were not as successful as in [76]. The number of iterations decreased at most by 15% due to the difference in the patterns of loading conditions.

1.2.3 Domain Partitioning

In a domain decomposition based parallel solution method, the performance of the solution is highly affected by the way the domain is partitioned into subdomains. The optimum partitioning for a particular domain involves many significant criteria such as balancing the workloads for each processor, minimizing the communication among the processors, having solvable data at each processor, and ensuring the parallel solution time to be less than the serial solution time. Moreover, at the same time, each partition should be optimized according to the parallel computer architecture or the properties of the communication network. All these variables increase the complexity of the partitioning problem; hence, it is the main reason for having several partitioning approaches that work well with a particular type of problem and a particular parallel computer architecture. These methods can be examined in two groups: static partitioning or mapping and dynamic partitioning which is sometimes referred to as dynamic-load-balancing.

1.2.3.1 Static Partitioning

Static partitioning methods are preferred when the workload and communication requirements are known before the computation starts. Hendrickson and Devine [16]

31

classified static partitioning methods in four groups: geometric, topological, graph based and hybrid methods. The hybrid method uses any of these approaches together. Simon et al. [43] considered optimization based techniques as another group. The optimization based methods view the partitioning problem as an optimization problem and try to solve it by applying cost function based optimization techniques such as genetic algorithms and neural networks.

Geometric Algorithms

Geometric based methods divide the domain by using the geometric properties of each object, i.e. nodal coordinates, elements etc. There are many geometric partitioning methods available in the literature but the basic ones are given below:

- **Recursive Coordinate Bisection Method (RCB):** The method is first proposed by Berger and Bokhari [2], which is based on the idea of recursively dividing 2D domains by lines and 3D domains by planes where the number of objects is equal to each other in each half. In this algorithm, cutting lines or planes are orthogonal to one of the coordinate axes.

- **Unbalanced Recursive Bisection Method (URB):** Jones and Plassmann [26], proposed a modified approach to the RCB method where instead of alternating the cut directions to force equal number vertices, their method chooses the cut in such a way that it will yield the smallest maximum aspect ratio of the resulting rectangles. In this way, they produced regions having better computational quality.

- **Recursive Inertial Bisection (RIB):** Simon [44] used a mechanical point of view and considered every vertex as point masses. Then, he calculated the direction of the principle inertial of the domain and the cutting plane was chosen to be orthogonal to it.

Topological Algorithms

Topological based methods use the connectivity information among the objects. Some of the topological methods are as follows:

- **Greedy Algorithm:** Greedy algorithm is a straight forward, fast, and simple decomposition algorithm developed by Farhat [9]. It uses the element connectivity information and tries to balance the number of elements or nodes and at the same time, minimize the interface nodes. It chooses a starting node and jumps to its neighboring node and continues to add nodes to the partition until the targeted number of nodes or elements are included. Then, starting from an interface node of the first partition, it starts creating the second partition using the same procedure. This procedure is repeated until the desired number of partitions is obtained.

- **Bandwidth Reduction Approach:** Malone [34] utilized a bandwidth reduction based partitioning approach in order to perform explicit linear integration. He first reordered the nodes so that the bandwidth of the connectivity matrix was minimized. Then, the elements were reordered by the sequence of their lowest numbered nodes. This element list was used to assign elements to the processors.

33

Finally, the interface nodes were processed to determine the most suitable processor to transfer the corresponding elements.

- **Octree Partitioning:** This method of partitioning was used together with an octree based mesh generator as a part of the research of Flaherty et al. [12]. In this approach, the problem is recursively divided into two in each coordinate direction until an application-specified number of octants are obtained. The octants are actually the leaves of the octree. Thus, the computational cost of each leaf is calculated depending on either the total number of elements or the total number of degrees of freedom. Since the total cost of the leaves and the number of partitions are known, each leaf is distributed to a partition provided that the cost does not exceed the optimal amount.

Graph Based Algorithms

In such methods, graph models of a computation are prepared and all the partitioning computations are performed on these models. Graph models are widely used to describe the data dependencies in the computation.

There are various types of graph models that exist in the literature. Hendrickson and Kolda [17] surveyed these graph models, their advantages, and deficiencies. Undirected graph model is widely used in most of the graph partitioning algorithms. It works efficiently if there is a symmetric data dependency and identical input and output data [17]. In order to overcome these shortcomings, non-standard models were proposed [4, 18]. For example, in the bipartite graph model developed by Hendrickson and Kolda [18], the vertices of a graph are divided into two disjoint subsets where the first set represented the rows of a matrix and the second one represented its columns. The edges cross between the rows and the columns. This model has the ability to represent unsymmetrical calculations and, moreover, allows that the initial partition to be different from the final one.

Another example of alternative graph model approach is the hypergraph. Proposed by Catalyurek and Aykanat [4], a hypergraph model assumes that edges can have more than two vertices. A hypergraph consists of a set of vertices and hyperedges contain a subset of vertices. They reported that, this kind of data dependency representation reduced the communication time by 30% when compared with the standard model.

There are many graph partitioning algorithms. The basic approaches are as follows:

- **Recursive Spectral Bisection Method (RSB):** This method uses the eigenvectors of a matrix associated with the graph called Laplacian matrix. Fiedler [11] has studied the properties of the second smallest eigenvalue λ and its eigenvector of the Laplacian matrix. He called λ the algebraic connectivity since it carried the information of the vertex and the edge connectivity of a graph. The differences between the eigenvector values represented the topological distance information among the vertices of a graph. He then investigated the partitions of a graph generated by the components of the second eigenvector.

  Pothen et al. [47] proposed a spectral algorithm that used the second smallest eigenvector of the Laplacian matrix for partitioning the vertices into two sets. They developed a bisection algorithm where they first computed the second median of the second eigenvector. The vertices having an eigenvector component

35

smaller than the median were collected in one partition and the rest were in the other partition.

- **Heuristic Partitioning Approaches:** Any direct approach which needs to find an optimal solution to the partitioning problem requires a huge amount of computations which might be very inefficient for several types of problems. Because of that reason, heuristic approaches are preferred since they can produce good solutions very quickly. Most of the present day heuristic partitioning algorithms are based on the study of Kerninghan and Lin [27]. Their approach started by dividing the partition into two. Then, they defined a cost function that considered the computation weight and the communication amount. By transferring elements from one partition to another, they tried to minimize the cost function.

1.2.3.2 Deficiencies of Static Partitioning Algorithms and Improvements
<u>Solution Time</u>

Depending on the size of the problem and the solution type, computing an eigenvector of a graph may be very computationally expensive to use. Barnard and Simon [1] implemented the multilevel approach to RSB algorithm in order to speed up the eigenvector computation. Multilevel algorithms based on the idea of coarsening the mesh by removing edges and vertices, hence, decreasing the size of the graph. By the application of a multilevel algorithm before the spectral decomposition, a considerable amount of speed-up was obtained without affecting the quality of the partitions significantly.

Hendrickson and Leland [21] utilized the same approach as Barnard and Simon [1]; however, they investigated the effect of transferring partitions instead of transferring eigenvectors. Their method is composed of three steps: coarsening, partitioning, and refinement. They preferred the spectral method for partitioning and the work of Fiduccia and Mattheyses [10] for refinement.

Multilevel methods are very popular since they decrease the size of the graph significantly. However, in order to have good quality partitions, the coarsening and refinement methods are very important.

A detailed investigation on the coarsening and refinement algorithm was done by Karypis and Kumar [33] for bisection based partitioning method. For the coarsening step, they compared four methods called: Random matching (RM), heavy edge matching (HEM), light edge matching (LEM), and heavy clique matching (HCM). Among all these methods, HEM performed the best in terms of the amount of edge-cut and solution time. Similar test runs were performed to check the efficiency of two refinement methods and both algorithms produced similar results.

Karypis and Kumar [32] performed a similar study for the multilevel k-way partitioning algorithm that they developed. They compared three different coarsening algorithms, random matching (RM), heavy edge matching (HEM), and modified heavy edge matching (MHEM) for the multilevel partitioning approach. Similar to the recursive bisection method, HEM produced the best results.

Bisection

When a graph will be partitioned into two, the eigenvector information provides the lowest edge-cut with balanced number of vertices in each partition. However, as the number of partitions increase, the eigenvector calculations must be repeated for each partition. This is a very expensive approach. Hendrickson and Leland [22] used multiple eigenvectors to allow partitioning to extend to four or eight subdomains at each stage of recursive decomposition. They developed a recursive spectral quadsection algorithm which decomposed the domain into 4^k partitions and similarly developed a spectral octasection algorithm which decomposed the domain into 8^k partitions.

Hsieh et al. [24] presented two RSB based methods that generalized the algorithm for arbitrary number of partitions. The first method was called the recursive sequential cut algorithm where the first subgraphs, having 1/p of the vertices, was cut from the domain where p is the number of subdomains. Then, from the remaining domain another subgraph is cut which had 1/(p-1) vertices. This process continued recursively until all subdomains were created. The second method, recursive spectral two-way algorithm, divided the graph into two parts of different sizes. The number of vertices in each subdomain was calculated by considering the number of vertices in other subdomains, total number of subdomains, and the total number of vertices.

Karypis and Kumar [32] presented a k-way partitioning algorithm where they directly partitioned the graph into k parts rather than recursively obtaining two-way partitioning of each, resulting partition. They used the multilevel approach where they first coarsened the graph, partitioned it into k parts, and refined it. In all these steps, their objective was to minimize the amount of edge-cuts. For the refinement step, they developed faster and equally effective methods called Greedy refinement (GR) and global Kernighan-Lin refinement (GKLR). Both methods produced the same amount of edge-cut; however, GR performed faster than GKLR.

Optimization

The most efficient partitioning methods attempt to balance the number of nodes or elements while minimizing the number of interface nodes. This idea works for many problems but when they are used together with a domain composition based iterative solver, the interface problem becomes difficult to solve as the aspect ratio of the subdomain increases. In order to overcome this difficulty, Farhat et al. [8] added a post optimization step for balancing the aspect ratios of the subdomains. In this study, they utilized both Greedy and recursive spectral bisection algorithm for partitioning. Then, they defined a cost function based on the aspect ratio, the size of the interface, and the load imbalance. By applying simulated annealing method, which is a technique to find a good solution to an optimization problem by trying random variations of the current solution, they improved the aspect ratios of the subdomains, hence, decreased the number of iterations of the FETI solution.

Another approach was the multi-constraint partitioning proposed by Karypis and Kumar [35]. Multi-constraint partitioning idea is useful for computations where the requirement of equal sized partitions only by itself is not enough. In such cases, the problem was considered in a more generalized way where the constraints were defined as a vector of weights assigned to each vertex. Hence, the aim of partitioning algorithm was to satisfy the balancing constraints associated with each vertex weight while trying to minimize the edge cut. Two different methods were developed to include the multiple

39

constraints, horizontal, and vertical formulation. The horizontal formulation was more suitable for the cases where both the computational and memory requirements are balanced. On the other hand, the vertical formulation was more appropriate for the multi-phase computations. For such computations, an explicit synchronization step is required at each computational phase.

Mixed Element Types

Another important deficiency of most of the partitioning methods when they are utilized with iterative solvers is that they may introduce mechanisms in the partitions. This situation often occurs in a 3D model which contains mixed dimensional elements. Day et al. [6] presented this problem in their study and showed that FETI method failed on every partition of the problem they were solving. They proposed a constraint for 2D and 3D elements that avoided mechanisms in each subdomain and called these subgraphs as "elastic connectivity graph". For 1D elements, instead of using beams, they recommended using multi-point constraints. Examples using elastic connectivity graph and multi-point constraints then executed without any problem.

Similarly, Topping [49] defended that the dual graph representation methods for finite element models were not applicable to the structures that consisted of both one and two dimensional elements. This was mainly because one-dimensional elements would give a zero dimensional mesh point in the graph. In order to overcome this difficulty, he proposed a new type of graph, a bubble graph, where not only the adjacency but also the geometrical arrangements of the elements were considered. He then partitioned three tower models consisting of 1D and 2D elements and presented results for dual, communication, and bubble graph representations. He concluded that this new approach

40

provide a better basis for partitioning, especially when additional elements were added to a current model.

Heterogeneous Environments

The other weakness of the current partitioning algorithms is that they all assume the same communication speed among the processors. As Hendrickson [19] pointed out, this was not the case in heterogeneous networks, especially for the clusters of symmetric multi-processors (SMP) where communication speed within an SMP was different than the speeds between SMPs.

Teresco et al. [48] focused on a similar problem. They attempted to create a scalable parallel adaptive technique for different parallel environments. They used available memory, available processing power, availability, bandwidth, and latency information of the communication resources to describe the properties of the parallel environment. Then, they developed a hierarchical partitioning method called the Rensselaer Partitioning Model for finite element meshes by providing additional information about the computational environment. They solved a parallel adaptive mesh computation on various parallel computers. They also tested several network connection types, slow and fast ethernet (10Mbit and 100Mbit) and non-blocking switch, on the same cluster of workstations. The test results showed that scalability for clusters of workstations was limited to 10 computers due to bandwidth and latency issues. After 8 processors, the performance of slow ethernet dropped very rapidly. The solution times with high ethernet were much better than the slow one. The non-blocking switch outperformed the ethernet connections since it allowed sending concurrent messages.

Pellegrini and Roman [45, 46] proposed the idea of dual partitioning. They created two graphs called the source graph and the target graph. The source graph consisted of vertices and edges assigned with the integer weights which define the computation weight and the communication volume. The second graph, called the target graph, modeled the target machine architecture. In this graph, the vertex weight represented the computational power of the corresponding processor and the edge weight represented the cost of the communication. The partitioning algorithm first partitioned the target graph by trying to keep the strongly connected clusters of processors together. Then the source graph was mapped onto the target graph by trying to minimize the communication as much as possible. When the partitioning was performed for homogeneous architectures, the quality of the partitions was similar to METIS [30] and CHACO [22]. When the partitioning performed for a hypercube machine, the 98% of edges of the resulting remained local.

1.2.3.3 Dynamic Partitioning

Some cases arise where it is either impossible to calculate the workload or the computational requirement varies over time in an unpredictable way. One approach to handle such a situation is to repartition the mesh by including the new information about the computational loads. However, it is very difficult to ensure that new partition will be close to the previous one. Otherwise, huge amounts of data must be transferred. The other method is to transfer nodes among the processors in order to balance the load by shifting the interfaces. The problem with this approach is that the shifting might cause an increase in the edge-cut which will increase the communication volume. Therefore, having a transfer algorithm that will balance the workload while keeping the edge cut as small as possible is very significant for such problems. This phenomenon is known as dynamic load balancing or dynamic partitioning in the literature.

Walshaw and Berzins [54] developed a migration strategy for the dynamic load balancing problem. This process can be broken down into two steps. The first step is called scheduling where an optimum schedule for each processor is set up in order to determine the exact amount of workload that a processor should send to or receive from its neighboring processors. This step is mostly iterative. For most of the applications a symbolic workload computation is performed without actually transferring vertices among the processors. Once the final schedule is obtained, each processor decides which vertex it should send to or receive from its neighboring processor. This step is called vertex migration.

In their research, Hu et al. [25] focused on the parallel finite element solution of a partial differential equation where the adaptive meshing technique was used. Hence at each step of the calculation, the computational load in each partition changed. In order to balance the work load, they proposed a dynamic load balancing method based on a diffusion algorithm. Their method was based on Cybenko's [5] approach where the problem was considered analogous to the movement of heat to reach equilibrium in an uneven temperature environment.

Schloegel et al. [40] developed a parallel partitioning and repartitioning library, PARMETIS [28] where they implemented various repartitioning algorithms. These algorithms were based on two different approaches. Their first approach was the scratch-remap algorithm based on the work by Oliker and Biswas [35]. In Oliker and Biswas's [35] approach, the imbalanced graph was partitioned from scratch using a multilevel graph partitioning algorithm. Then, the new partitions were intelligently mapped to the original partitions in order to reduce the amount of vertex migration. PARMETIS [28] also implemented the multilevel version of the scratch-remap algorithm.

The second approach was the diffusion approach. PARMETIS [28] implemented two different types of diffusion algorithms, local and global. For both algorithms, a parallel coarsening was first performed in order to decrease the size of the graph. During coarsening, only the vertices belonging to the same partition was considered for merging. This way, the initial partition of the coarsest level graph became identical to the input partition that was being repartitioned. Then, scheduling started. In the global diffusion approach, the scheduling was performed according to the Hu et al.'s [25] method. In the local diffusion approach by Schloegel et al. [41], the vertex migration decisions were made at every partition according to the relative difference in partition weights between each partition and all of its neighbor partitions. Once the scheduling step ended, refinement began. Each vertex was visited randomly and checked whether migration to another partition would help balancing the weights while keeping the changes in partitions and the edge-cut as minimum as possible.

Van Driessche and Roose [7] introduced virtual edges and virtual vertices where the weight of a virtual edge is equal to the cost of transferring the corresponding grid to another processor. Then, they calculated the Fiedler vector of the extended graph and decided on how to divide graph into two using the virtual vertices.

By using the existing dynamic optimization methods, Walshaw and Cross [52] developed three different optimization algorithms to be used together with the multilevel graph partitioning techniques. The first method created independent optimization

44

problems at the interface regions and then solved them. The second method was a derivative of the first one where vertices can migrate only in one direction. The third method used the relative gain approach while selecting the appropriate vertices to migrate. These methods were then compared with other partitioning tool, PARMETIS [28]. The results of the above algorithms produced mostly better quality partitions; however, the computations required more time than the PARMETIS [28] method.

Although the spectral bisection method produces very good quality partitions, it is not suitable if the computation requires repeated partitioning. In order to overcome this difficulty, Simon et al. [43], extended the spectral algorithm to handle the dynamic load balancing. For this purpose, a spectral basis was calculated using the eigenvectors of the coarsest mesh. During the computation, when a change in a partition occurs, new vertex weights which corresponded to the changed computational load were calculated. Then, the graph was repartitioned with recursive inertial bisection method in the spectral coordinates.

A similar approach is sometimes used for improving the quality of the partitions. Vanderstraeten and Keunings [51] developed a two-step partitioning to improve the workload distribution among the partitions. Their aim was to obtain a sub-optimal solution by using a heuristic approach since the solution complexity of graph partitioning problems grew exponentially with the problem size. They initially partitioned the graph with recursive graph partitioning technique. Then, they introduced a cost function that described the load imbalance among the processors. By using three different heuristic methods, namely 'Simulated Annealing', 'Stochastic Evolution', and 'Tabu Search', they balanced the loads among the processors.

In another study by Vanderstraeten et al. [51], the two-step partitioning algorithm was tested using the Greedy algorithm and multilevel recursive spectral bisection method. This time, the cost function was based on three control functions defining the interface size, load imbalance, and subdomain aspect ratio. Then, they benchmarked the partitioning algorithm, with or without the optimization step, together with the various solution algorithms. They considered a domain decomposition-based iterative algorithm and the frontal method for the solution of sparse linear equations, and explicit time integration. For most cases the optimization step decreased the total solution time for both the Greedy algorithm and multilevel recursive spectral bisection method.

1.2.4 Condensation Algorithms

Condensation is a process of reducing the number of degrees of freedom by substitution. There are primarily two approaches. In the direct approach, the direct decomposition of the equations stops before the stiffness matrix has been fully reduced. The details of this approach are very well explained in many finite element texts such as the one by Cook et al. [66]. The other approach, sometimes referred as the dual condensation method, is an iterative approach where the interface unknowns are chosen as Lagrange multipliers representing the surface tractions at the interface boundaries [76].

Wilson and Dovey [101] developed a direct solution algorithm that can be used to form the condensed stiffness matrix of a substructure. It used the active-column matrix storage scheme. Moreover, the algorithm was designed in such a way that it allows out-of-core storage. The LU decomposition method was utilized for solution.

Han and Abel [86] modified the usual matrix decomposition approach in order to decrease the number of operations. Their method was based on Cholesky decomposition. They reformulated the decomposition method to reduce the number of multiplications. Moreover, the method allowed using the active column scheme and out-of-core storage.

1.2.5 Overview

In the literature, there are several parallel solution methods that have a global [55, 57, 70, 88] or domain-by-domain [62, 73, 80, 82, 89] approach with direct or iterative solvers. However, their performance may be limited depending on the type of the analysis, the parallel environment and the structural properties of the system. For example, iterative solvers are not suitable for problems with multiple loading conditions since they must start the solution from scratch for every loading condition. Global solution approaches focus on speeding-up the solution of the linear equations only. They generally work with a factorization schedule that affects the way the elements of the stiffness matrix are distributed among processors. Thus, the effectiveness of such methods also depends on the efficient implementation of the stiffness and the force matrix generation, the distribution of the stiffness matrix, force matrix, and the nodal displacements among processors, and element force computations. Moreover, global solvers need efficient out-of-core solution approaches for large structures.

Among all these solution methods, substructure based methods offer several advantages. They have the capability of parallelizing every step of the solution, from element stiffness generation to element result computations. They minimize the communication requirement by performing the parallel solution at the substructure interfaces which makes them very suitable for parallel environments with low

47

communication performance. Moreover, if direct solvers are utilized, they will be robust and become very suitable for problems with multiple loading conditions.

The efficiency of the parallel substructure based solution methods highly depends on the way the structure is partitioned into substructures. Existing partitioning algorithms attempt to keep the number of interface nodes as small as possible while keeping a balanced number of nodes or elements in each substructure [17]. Unfortunately, this partitioning approach does not create substructures that have balanced condensation times [19]. As a result, some processors may stay idle during condensation which significantly decreases the performance of the solution. Only a few studies have focused on this problem [71, 82, 102]. These past studies are either for shared memory architectures or the proposed workload balancing algorithms are too slow to be utilized prior to linear static solution. Currently, a fast and effective method that balances the condensation times of the substructures for distributed memory architectures has not yet been developed.

Another important problem of the substructure based solution approaches is the interface solution. Past studies have used parallel active-column solvers [80, 89]. Such solvers required the distribution of the factorized interface stiffness matrix among processors during back substitution which significantly decreased the performance of the solution for distributed memory architectures. Because of that reason, Farhat et al. [80] recommended performing forward and back substitutions serially. Likewise, Yang and Hsieh [102] utilized serial algorithms for interface solution in their recent work. The serial portions of the interface solution method decrease the scalability of the substructure based methods.

1.3 Objectives and Scope

The main purpose of this study is to develop an efficient parallel solution algorithm for linear systems having multiple loading conditions. The ultimate goal is to demonstrate an approach that can be used by structural engineers to efficiently analyze large linear structural problems.

Thus, the research objectives are as follows:

- Developing a complete parallel substructure based solution framework, from partitioning to element result computations so that the framework will be ready to be utilized by the structural engineers. Moreover, by investigating all of the steps of the parallel solution such as as: partitioning, equation numbering, workload balancing, condensation, and the interface solution, the complications that arise from combining these separate steps will be observed and improved.

- Developing a workload balancing method that minimizes the condensation time differences between the substructures and thus decreases the total solution time.

- Designing an interface solution algorithm specifically for the linear static solution with multiple loading conditions.

- Developing solution algorithms that are the most applicable for PC clusters. This way, many design offices can utilize their existing hardware to perform parallel computation.

As mentioned previously, this research focuses on the solution of linear systems with multiple loading conditions. The solution algorithm is a parallel substructure solution method with LU decomposition method based solvers. The target parallel environment is a PC clusters system connected using a network hub, a router, or a switch. Currently, only homogenous PC clusters will be considered but the structure of the solution algorithm will be designed in such a way that it can be utilized in heterogeneous computing environments with minor modifications.

A complete parallel solution framework for the analysis from partitioning to element result computation is developed. The framework includes the workload balancing step that is fast enough to be utilized prior to linear static analysis. The MPICH [115] library is used for the communication and the data exchange between the processors. The database has an object-oriented structure and the design will be performed using the UML [136] (Unified Modeling Language) modeling language. The program is designed to work under the Windows operating system.

The example problems focus on the solution of actual civil engineering structures composed of mixed structures, 1D elements mixed with 2D or 3D elements with multiple loading conditions. The performance of the partitioning, equation numbering, workload balancing, condensation, interface, and global solution algorithms are investigated.

1.4 Thesis Outline

The remainder of the thesis is organized as follows. Chapter 2 presents an overview of the existing partitioning and repartitioning algorithms. Then, specific attention is devoted to the METIS [30] and PARMETIS [28] partitioning and repartitioning algorithms. Chapter 3 presents the active-column direct condensation algorithm that was utilized during the local solution of substructures. In this chapter, the computation speed of the condensation algorithm, condensation time estimation, and out-of-core solution are discussed. Chapter 4 presents the workload balancing method developed to decrease the estimated condensation time imbalances among the substructures. The effectiveness of this method will be demonstrated for idealized example problems and actual structural models. The parallel variable band solver that is utilized for the interface solution is investigated in detail in Chapter 5. Its performance on three different PC Clusters is examined and compared.

Chapter 6 combines all the algorithms and the findings of the previous chapters and presents the implementation of the substructure based parallel solution framework. Then, the complete parallel solution results for the selected example problems are presented.

The final chapter summarizes the important results of this study and concludes with recommendations for future work.

CHAPTER 2

INITIAL PARTITIONING AND REPARTITIONING

2.1 Introduction

The first step of a substructure based parallel solution method is partitioning the structure into a number of substructures. The goal of the partitioning is generally to balance the computational workloads of each processor while keeping the size of the substructure interfaces as low as possible. When the processor workloads of processors are balanced, the processors would be more efficiently utilized. In other words, none of the processors will stay idle while waiting for other processors to finalize their computations. When the size of the substructure interfaces is low, less time will be spent for data transfers among substructures. Thus, the ultimate goal of partitioning is actually to decrease the computation time.

There are many different partitioning approaches in the literature, but they can be examined in two main categories: static and dynamic partitioning. Static partitioning algorithms are mostly utilized in problems where the workload is computable before the solution and remains unchanged during the solution. For such problems, the computational workload is usually represented as a single integer value assigned to the nodes or the elements of a structure. Hence, once the sum of the weight values of each substructure are balanced, it is assumed that the computational workloads for each processor are also balanced.

The dynamic partitioning algorithms, on the other hand, are developed for problems in which the computational loads can not be known prior to partitioning or the computational loads of processors change during the solution. Dynamic partitioning algorithms mainly modify the substructures according to the new computational loads in such a way that the loads are balanced, the differences between the previous and newly formed substructures are minimized, and the interface size of the new substructures has not significantly increased when compared with the interface size of the previous substructures.

In this study, static partitioning was utilized for the initial partitioning of substructures prior to the parallel solution. Dynamic partitioning was then used to repartition the structure in order to balance the workload condensation times. A brief discussion of static and dynamic partitioning follows.

2.2 Initial Partitioning (Static Partitioning)

Static partitioning algorithms partition a structure into a desired number of substructures by using various approaches. The different partitioning approaches can be classified in three groups:

1. Geometric methods

2. Topological methods

3. Graph methods

The geometric methods use the geometrical information of a structure. Such methods commonly use the coordinates of the nodes in order to find a good bisection line that divides the structure into two. Then, the new partitions are divided into two with the bisection lines. This procedure continues recursively until the desired number of partitions is created. These methods are very fast methods but the interface size of the partitions are much larger when compared with other partitioning methods [16].

Topological methods, on the other hand, use the connectivity information of the elements in a structure. One of the most well-known topological methods, the Greedy algorithm [9], begins partitioning by choosing a starting node. It extends the partition by adding the neighboring nodes of the starting node. Then, it adds the neighboring nodes of newly added nodes. The same procedure is repeated until the correct number of nodes has been included. Construction of the next partition begins from the boundary of the previous partition and a similar procedure is followed until the new subdomain is created. This continues until the whole domain is decomposed. It is a simple and fast partitioning method and the partitions generally yield to a reasonable number of interface nodes [9].

The graph methods work with the graph representation of a structure which describes a structure in terms of vertices and edges. Each vertex is actually a solution point and its weight shows the computational weight of that point. An edge is used to define interactions between the vertices. Therefore, a graph partitioning algorithm attempts to keep the vertex weights balanced in each partition while keeping the edges at the domain interfaces as small as possible. One approach used to accomplish this is to create a mathematical definition of the partitioning problem and attempt to solve it. In the literature, this mathematical definition is an NP-complete problem [7, 13, 21, 24] which has a computable solution. However, the exact solution could be very expensive for some types of the analysis methods since the exact solution requires the computation of the second smallest eigenvector of the system.

The other approach is to target a reasonably good solution instead of the best one with the application of various heuristic methods. The multilevel scheme is a commonly used approach where the size of the graph is reduced and partitioning is performed on a relatively smaller graph. Although the partition quality decreases due to coarsening, the partitioning time drops considerably.

Karypis and Kumar [32] performed a comparative study on various static partitioning algorithms. They investigated the algorithms in terms of partition quality, local and global view, run time, and degree of parallelism as shown in Figure 2.1. The partition quality refers to the number of edges at the partition interfaces (edge-cut). If the edge-cut is high, there will be more communication required during the solution. Local and global view properties of the algorithms affect the quality of partitions. The local view indicates whether the algorithm is able to perform localized refinement. The global view refers to the extent that the partitioning algorithm takes into account the structure of the graph. Each square in the run-time indicates a factor of 10 increase in run-time when compared with the coordinate nested dissection algorithm. The degree of parallelism indicates if the parallel implementation of the algorithm is possible.

	Number of Trials	Needs Coordinates	Quality	Local View	Global View	Run Time	Degree of Parallelism
Spectral Bisection	1	no	●●●●	○	●●●●	■■■■	▲▲
Multilevel Spectral Bisection	1	no	●●●●	○	●●●●	■■■	▲▲
Mulitlevel Spectral Bisection-KL	1	no	●●●●●●	●●	●●●●	■■■	▲▲
Multilevel Partitioning (including METIS)	1	no	●●●●●●	●●	●●●●	■■	▲▲
Levelized Nested Dissection	1	no	●●	○	●●	■■	▲▲
Kernighan-Lin	1	no	●●	●●	○	■■	▲
	10	no	●●●◐	●●	●◐	■■■	▲▲
	50	no	●●●●	●●	●●	■■■■◐	▲▲
Coordinate Nested Dissection	1	yes	●	○	●.	■	▲▲▲
Inertial	1	yes	●●	○	●●	■	▲▲▲
Inertial-KL	1	yes	●●●●	●●	●●	■■	▲
Geometric Partitioning	1	yes	●●	○	●●	■	▲▲▲
	10	yes	●●●◐	○	●●●◐	■■	▲▲▲
	50	yes	●●●●	○	●●●●	■■■◐	▲▲▲
Geometric Partitioning-KL	1	yes	●●●●	●●	●●	■■	▲
	10	yes	●●●●●◐	●●	●●●◐	■■■	▲▲
	50	yes	●●●●●●	●●	●●●●	■■■■◐	▲▲

Legend

●	10% improve in edge-cut
■	10 times increase
▲	Serial
▲▲	Moderate
▲▲▲	High

Figure 2.1 Characteristics of Various Partitioning Algorithms [33]

The spectral bisection algorithms are graph based methods which perform partitioning by computing the second smallest eigenvector of the system. The inertial methods consider every vertex as a point mass and compute the graph's principal axis. Then, they divide the graph into two by using a line that is orthogonal to the principal axis. Multilevel partitioning and Kerninghan-Lin are heuristic methods. Among all these methods, the multilevel partitioning approach produced the best quality partitions at a relatively faster speed when compared with other approaches that produce good quality partitions. For this reason, METIS [30], a multilevel graph partitioning library, was chosen to perform initial partitioning in this study.

2.2.1 METIS Library

METIS [30] is a software package developed for partitioning large irregular graphs. It utilizes a multilevel approach to speed-up the partitioning process and allows single or multiple vertex and edge weight definitions.

METIS [30] works with the graph representation of a structure and accepts either a nodal or a dual graph. An example of a nodal graph for an arbitrary structure (Figure 2.2) is given in Figure 2.3a. In the nodal graph, each node corresponds to a vertex in the graph. The vertices are joined with an edge if the corresponding nodes are connected by an element. On the other hand, in the dual graph (Figure 2.3b), each element of the mesh corresponds to a vertex in the graph. The vertices are connected with an edge if the corresponding elements in a mesh share a face or an edge.

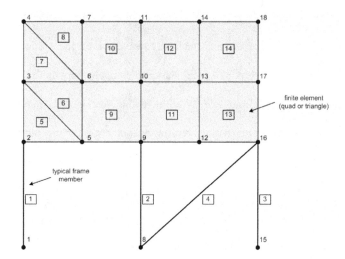

Figure 2.2 An Arbitrary Structure

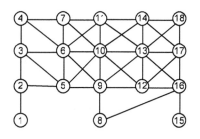

(a) Nodal Graph

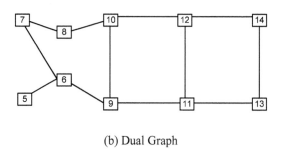

(b) Dual Graph

Figure 2.3 Graph Representations of an Arbitrary Structure

58

In a dual graph, a finite element of dimension 'n' has boundaries of dimension '(n-1)'. For a 1D element (frame element), the boundary has zero dimension. That's why, a special algorithm is necessary to generate a dual graph [49] for structures having frame elements.

METIS [30] partitioning algorithms are based on the multilevel approach. As illustrated in Figure 2.4, multilevel approaches are composed of three phases: coarsening, initial partitioning, and refinement

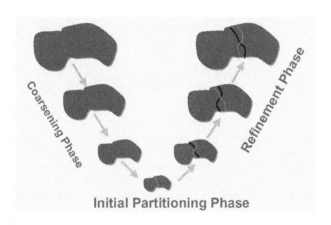

Figure 2.4 Multilevel Approach [31]

2.2.1.1 Coarsening Phase

The coarsening phase is basically the construction of a sequence of smaller graphs, each having fewer vertices from the original graph. METIS [30] uses two different approaches to obtain a coarser graph. The first approach is based on finding a random matching (Figure 2.5) of vertex couples and collapsing the matched vertices into a multinode. A matching is defined as a set of edges, not two of which are incident on the same vertex. Thus, the next level of coarser graph is constructed from the finer graph by first finding a matching. Then the matched vertices are joined and a multinode is created. The vertex weights are added and assigned to a newly created multinode. The advantage of this approach is that the coarsened graph preserves many properties of the original graph.

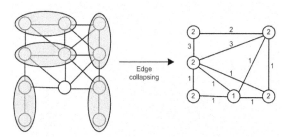

Figure 2.5 Matching and Edge Collapsing

The other approach is based on creating multinodes that are made of groups of vertices that are highly connected. This approach is most suitable for the graphs that have various and high connection patterns. However, fine graphs arising from FE applications have mostly similar connection patterns; therefore, the first approach is best suited for such graphs.

2.2.1.2 Initial Partitioning Phase

METIS [30] uses two approaches for the partitioning of a coarsened graph. The first approach, called multilevel recursive bisection, is to perform log(k) levels of recursive bisection. In other words, the desired number of partitions (k) is obtained by dividing the graph and the subsequent subgraphs into two. The algorithm initially partitions the coarser graph into two. During the partitioning, the algorithm computes a high quality bisection of the coarser graph in such a way that each part has a predefined percentage of the total vertex weights. Next, each partition is refined. The same procedure is repeated on newly obtained subgraphs until the desired number of partitions is obtained.

The other approach, called multilevel k-way partitioning, directly partitions the graph into 'k' parts during the initial partitioning phase. In this algorithm, the graph is coarsened and refined only once. However, a k-way refinement algorithm is required to improve the quality of the partitions.

The recursive partitioning algorithm produces better quality partitions than the k-way partitioning algorithm. However, as the number of partitions increases (>32), the recursive partitioning algorithm becomes considerably slower than the k-way partitioning algorithm [32].

2.2.1.3 Refinement Phase

The final phase of multilevel partitioning is the refinement phase where the partitions of the coarser graph are projected back to the original graph by going through finer and finer graphs. Since finer graphs have more vertices, the partitions can still be improved by transferring vertices at the interfaces with local refinement heuristics.

It is reported in many studies that the local refinement algorithms based on Kerninghan-Lin [27] heuristic produce very good results for bisections. They basically swap vertices between partitions at the bisection to reduce the edge-cut until the movement of any vertex from one partition to the other does not improve the edge-cut.

On the other hand, refining a k-way partitioned graph is rather more complicated. In such a case, the vertices can move from one partition to many others. The METIS library uses a simplified version of k-way Kerninghan-Lin algorithm [32] for that purpose.

2.3 Repartitioning (Dynamic Partitioning)

Many of the static partitioning algorithms try to balance the vertex weights in each partition while keeping the communication volume at a minimum by reducing the total number of edges cut by the partitions (edge-cut). This approach works fine if the computational loads of a node or element can be represented as a vertex weight. However, there is a class of problems where either the computational load changes as the solution proceeds or it is not possible to compute the computational load of partitions before partitioning. For such cases, repartitioning algorithms must be utilized.

There are many different types of repartitioning algorithms but they can be classified into two groups according to their use of the original partition. The first group of algorithms, e.g. scratch-remap, first partition the graph from scratch according to the new vertex weights and use the existing partitioning information to minimize the difference between the original and the new partitions. The other group of algorithms first calculates the imbalance of the original partitions and then attempts to balance them by migrating vertices from the overweight partitions to the under-weight ones. Diffusion algorithms belong to this group.

2.3.1 Scratch-Remap Algorithms

The scratch-remap partitioners start with a newly computed partitioning and try to minimize its difference from the original partitioning. Simon et al. [43] utilized spectral coordinates for that purpose. They first calculated a spectral basis from the eigenvectors of the initial graph at the coarsest level. When the computational loads were changed, new vertex weights were computed that reflect these changes. Then, the new partitioning was computed by using a recursive inertial bisection method in the spectral coordinates.

Oliker and Biswas [35] developed a scratch-remap based load balancing framework for adaptive meshes. In their method, the imbalanced graph was partitioned from scratch using a multilevel graph partitioning algorithm. Then, the new partitions were intelligently mapped to the original partitions in order to reduce the amount of vertex migration. In another study [3], they tested various remapping approaches. The first approach (TOTALV) minimized the total volume of data moved among the processors whereas the second approach targeted (MAXV) minimizing the maximum flow of data from or to a single processor. The TOTALV approach targeted minimizing the remapping time by reducing the network contention and the total number of elements moved. On the other hand, the MAXV approach was more suitable for problems where it was more important to minimize the workload of the heavily-loaded partition than to minimize the sum of all loads.

According to Schloegel et al. [39], partition remapping is a three-step process. It needs original and newly computed partitioning information. First, a similarity matrix is constructed whose rows represent the domains of the original partitioning and columns represent the domains of the new partitioning. The element (i,j) shows the sum of the sizes of vertices that are in domain i of the original partitioning and in domain j of the new partitioning. In order to minimize TOTALV, the elements of the similarity matrix are selected in such a way that every row and column contains exactly one selected

63

element and the sum of their sizes is maximized. Then, for each element (i,j), the partition i is renamed to partition j on the remapped partitioning. As a result, the amount of overlapping between the original and the new partitioning is maximized.

2.3.2 Diffusion Algorithms

Diffusion repartitioning algorithms consider the problem analogous to the diffusion process where an initial uneven temperature in space drives the movement of heat, and eventually reaches equilibrium [25]. There are several algorithms based on this analogy. Cybenko [5], suggested an iterative algorithm where the processors exchanged load with their neighbors in such a way that the amount is proportional to the difference in their loads. He grouped the processors in pairs and the loads were exchanged between the pairs, thus decreasing the number of iterations. Ou and Ranka [36] developed a method that minimized the one-norm of the diffusion solution. By using the results, they calculated a solution vector for the movement of necessary vertex weights. Then, the graph was refined in order to decrease the edge cut. Hu and Blake [25], on the other hand, calculated the diffusion solution by minimizing the Euclidian-norm. Walshaw and Berzins [54] examined the repartitioning problem in two steps: scheduling and migration. The scheduling step involved the decision about the amount of load for each processor that should be send to or receive from neighboring processors. The migration step symbolically redistributed the vertices without increasing the edge-cut. Walshaw and Berzins thus [54] implemented Hu and Blake's method [25] in the JOSTLE [52, 53, 54] library for the scheduling step. Then, Walshaw and Cross [52] implemented various migration techniques in the JOSTLE [52] library.

In Hu and Blake's proposed scheduling approach [25], the average load per processor is first calculated using Equation (2.1):

$$\bar{l} = \frac{\sum_{i=1}^{p} l_i}{p} \qquad (2.1)$$

where l_i is the load of each processor (sum of the vertex weights) and p is the number of processors. Then a directional scalar variable, δ_{ij}, which shows the amount of load to be sent from processor i to processor j, is associated with each edge (i,j) of the graph. If it has a positive value, it means that the processor i will send the amount of δ_{ij} load to processor j.

The load balancing schedule should make the load on each processor equal to the average load, that is,

$$\sum \delta_{ij} = l_i - \bar{l} = 0 \qquad (2.2)$$

Equation (2.2) has infinitely many solutions, thus the solution which minimizes the data movement should be chosen. In order to accomplish this, let A be the matrix associated with Equation (2.2), x is the vector of δ_{ij} and b is the right-hand side. Then, the problem becomes the minimization of the Euclidian-norm of x subjected to A, Equation (2.3)

$$\text{Minimize} \quad \frac{1}{2}x^T x$$
$$\text{subject to} \quad A \cdot x = b \qquad (2.3)$$

65

where

$$A_{ij} = \begin{cases} 1 & i > j \\ -1 & i < j \\ 0 & i = j \end{cases}$$

Applying the necessary condition for the constrained optimization, Equation (2.3) becomes

$$x = A^T \lambda \tag{2.4}$$

where λ contains the Lagrange multipliers. Substituting back into (2.2) gives

$$L\lambda = b \tag{2.5}$$

with $L = AA^T$

Thus, once the Lagrange vector from (2.5) has been calculated, the amount of load to be transferred from processor i to processor j is equal to $\lambda_i - \lambda_j$ where λ_i and λ_j are the Lagrange multipliers associated with processors i and j, respectively, according to Equation (2.4).

Schloegel et al. [41] named the diffusion algorithms that used a scheduling similar to Hu and Blake [25] as "directed" or "global" diffusion algorithms since the diffusion is guided by the global picture of the partitions (Euclidian-norm). There are also "undirected" or "local" diffusion algorithms which occur through the distributed actions employing only the local views of the graph. Thus, vertex migration decisions are made at every partition according to the relative difference in partition weights between each partition and all of its neighboring partitions [41].

The migration step depends on the type of scheduling. In case of local diffusion, Schloegel at al. [40] utilized the following algorithm. First, the interface vertices are visited in a random order. If a vertex belongs to an overweight partition, it will be moved to an adjacent partition with lower weight. If there is more than one adjacent partition satisfying this condition, the one that leads to the smaller edge-cut will be selected. If a vertex belongs to an average-weight partition, it will be moved to a partition that leads to a reduction in the edge-cut as long as it does not make the destination partition overweight. If a vertex belongs to an under-weight domain, it will not be moved. This process is repeated until either the balance is obtained or no progress is made in balancing.

In the global diffusion algorithm by Schloegel et al. [40], the results of the Hu and Blake's method [25] are utilized. The results are represented as a transfer matrix whose element (i,j) shows how much load the processor 'i' needs to transfer to the processor 'j'. Again, the interface vertices are visited in random order. If a vertex has a neighboring partition which according to the transfer matrix needs workload, the vertex can be migrated to the neighboring partition. If a vertex is a neighbor of more than one such domain, it will be migrated to the domain that produces the highest gain in edge-cut. After a vertex is migrated, the transfer matrix is updated to reflect the vertex migration. After each interface vertex is visited once, the process is repeated until either the balance is obtained or no progress is made in balancing.

Both algorithms start at the interface nodes and gradually move into the mesh. In other words, they try to find a balanced partitioning by shifting the interfaces of the partitions.

2.3.3 PARMETIS Library

PARMETIS [28] is a parallel multilevel graph partitioning and repartitioning library that implements both scratch-remap and diffusion based repartitioning algorithms. It uses a multilevel scheme to speed-up the partitioning process. Moreover, it performs all the computations in parallel that enables it to partition and repartition very large sized graphs in a very short amount of time.

PARMETIS [28] implemented both types of diffusion algorithms, local and global. For both cases, it first performs a parallel coarsening to the partitions. In this case, only pairs of nodes that belong to the same partition are considered for merging. This way, the initial partition of the coarsest level graph becomes identical to the input partition of the graph that is being repartitioned. Hence, the coarsening phase will be local to each processor and easily parallelized. Then, in case of global diffusion, the scheduling is computed for the coarsest graph and vertex migration is performed. However, it is possible not to obtain a balance in the partitions at the coarsest level. If this is the case, the graph is uncoarsened one level to increase the number of vertices. Then, the scheduling and migration process is repeated for this level of the graph. This process continues until the partitions are balanced.

In the parallel version of the global diffusion algorithm, the scheduling computations are performed serially. Since the global diffusion algorithm is inherently random, each processor simultaneously computes a balanced solution. The balanced solutions will be different from each other; thus the partitioning solution that produced the lowest edge-cut is selected. The scheduling computations are performed once on the coarsest graph. If the partitions are not balanced, parallel multilevel local diffusion algorithm is utilized to balance the minor imbalances at the finer levels.

When the multilevel diffusion ends, the multilevel refinement begins on the current graph. Each vertex is visited randomly. Each interface vertex is checked whether migration to another vertex will satisfy the refinement migration criteria. This criterion aims to improve the edge-cut and is composed of three rules:

- Maintain the edge-cut, maintain the balance, and the selected partition is the vertex's initial partition.

- Decrease the edge-cut while maintaining the balance.

- Maintain the edge-cut and improve graph balance.

The multilevel refinement continues until the original graph is extracted.

If the local diffusion is preferred, the algorithm described in the previous section for the local diffusion will be utilized on the coarsest graph. Similarly, if the balance is not obtained, the graph will be uncoarsened to one level and the same procedure will be applied. This procedure continues until the partitions are balanced.

The parallel implementation of the local diffusion algorithm consists of two phases. During the first phase, the vertices are migrated only from the lower to higher numbered processors. During the second phase, vertices are migrated from the higher to lower numbered processors. This way, any unexpected edge-cut increases caused by the simultaneous migration of neighboring vertices are avoided. At each phase, the vertices are visited and selected for migration according to the local diffusion algorithm described in the previous section.

The parallel multilevel refinement is very similar to the parallel multilevel local diffusion algorithm. In this case, the vertices are moved according to the refinement migration criteria.

The parallel multilevel diffusion algorithms of PARMETIS [28] library are made up of three phases: graph coarsening, multilevel diffusion, and multilevel refinement as shown in Figure 2.6. In case of local diffusion, all three phases are performed in parallel whereas in the case of global diffusion, coarsening and refinement are performed in parallel, but the diffusion is performed serially on all computers.

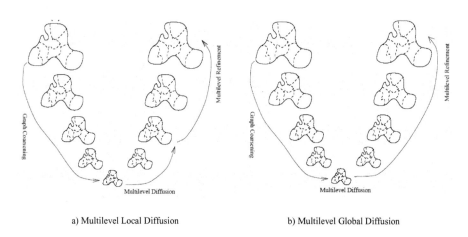

a) Multilevel Local Diffusion   b) Multilevel Global Diffusion

Figure 2.6 Multilevel Diffusion Algorithms a) Local b) Global [42]

PARMETIS [28] also implemented two parallel versions of the scratch-remap repartitioning algorithms. The first method is the parallel implementation of the method developed by Oliker and Biswas [35]. The original graph is first partitioned from scratch by using parallel multilevel k-way partitioning algorithm [42]. Then, the next step is remapping. Remapping is performed on the fine graph. Each processor computes one row of the similarity matrix simultaneously based on its current partition and the new partitioning. This information is distributed to a single processor and the similarity matrix is assembled. Then the new partition to processor mapping is computed and distributed to the other processors.

The second version of the parallel scratch-remap algorithm is based on multilevel approach. The coarsening is performed locally; in other words, the matchings are restricted to be among the vertices that are on the same partition. The result is that vertices of each successively coarser graph correspond to regions within the same domain of the original partitioning. Moreover, the edges of the original partitioning still remain visible even in the coarsest graph. Hence, in those portions of the graph that are relatively undisturbed by adaptation, the initial partitioning algorithm will have a tendency to select the same partition boundaries.

Once the coarsest graph is obtained, initial partitioning is performed. The remapping is applied on the coarsest graph by utilizing the method developed by Oliker and Biswas [35]. Then, the multilevel refinement begins. It uses three criteria into consideration when determining whether to migrate vertices. The refinement algorithm visits the interface vertices one-by-one and migrates them if they satisfy the following criteria:

- Decrease the edge-cut while still satisfying the balance constraint.

- Decrease the TOTALV while maintaining the edge-cut and still satisfy the balance constraint.

- Improve the balance while maintaining the edge-cut and the TOTALV

This refinement approach not only focuses on reducing the edge-cut but also reduces TOTALV while maintaining the balance of partitions.

CHAPTER 3

CONDENSATION

3.1 Introduction

The condensation algorithm plays a significant role on the effectiveness of the parallel solution since most of the total solution time is usually spent during condensation. Moreover, the selection of either an iterative or direct solver and a sparse or active column storage scheme affects the way the substructures are created. Thus, deciding on the condensation method has a key role on many of the algorithms utilized in this framework.

There are various condensation methods in the literature [66, 86, 76]. In terms of solution approaches, they can be classified as iterative and direct methods. Farhat et al. [75, 76], preferred to use iterative solution approach to solve the local problem at substructures in the FETI method in their earlier works since it required less storage space and was very well suited when combined with their iterative global solution algorithm. However, as the second generation of FETI methods were evolving, researchers began using direct methods due to their robustness [73]. Furthermore, direct methods require neither the computation of all the zero energy modes of the substructures nor developing techniques to prevent the zero energy modes from occurring.

Iterative methods are not suitable for problems having multiple right hand sides by nature. For each loading, the solution must start from the beginning. Although, there are a couple of proposed methods [76, 95] to improve the convergence of iterative methods for repetitive loadings, they provide very little improvement in performance when the consecutive loadings are not similar [62].

In this study, the direct condensation method was chosen due to the following reasons:

- **Robustness**: Direct condensation methods can handle all types of substructure shapes. Not only does it not require mechanism free partitions, but it also can handle distinct structures in a single substructure.

- **Predictability**: Once the structure is partitioned, it is possible to calculate the amount of time required for condensation. Similarly, the time required to factorize a single load case is also computable. This property is very useful during workload balancing.

- **Reusability:** The equations are factored only once when direct condensation method is used. Once the factorization coefficients are calculated, they are used to factorize the load vectors. This property is very suitable when there are multiple loading conditions.

The direct condensation algorithms differ according to their matrix storage schemes. There are various parallel solution studies which use sparse solvers with compressed row matrix storage scheme [59, 102] and others that are based on an active column storage scheme [63, 72, 80]. Chuang and Fulton [64] performed a comparative study for the sparse and active column solvers in a parallel environment having both local and shared memory. They concluded that the efficiency of either solver was problem dependent and mainly depended on the sparsity of the stiffness matrices. Farhat et al. [73] also used both active column and sparse solvers for the solution of the local problem of the FETI method. They presented two example problems where both solvers outperformed the other in one of the problems. Moreover, they showed that the sparse solver was slower than the active column solver during forward and backward substitutions due to inefficient cache utilization. Thus, it has not yet been shown the superiority of one solver to another especially for the substructure based parallel solution methods. Furthermore, each solver's performance on ordinary PCs which have relatively low cache memory and a slow bus speed is questionable.

In this study, an active column based condensation algorithm was utilized to reduce the degrees of freedom of substructures. Its superiority in load factorization is a major advantage for this framework.

3.2 Method

3.2.1 Theory

For the condensation algorithm, the method presented by Wilson and Dovey [101] was utilized in this study. Their method is based on LU Decomposition and uses active column storage. The algorithm assembles the interface equations at the end of the matrix for condensation. A common approach [60, 63, 81, 89] is to first renumber the internal equations by using some type of a profile minimization algorithm. The interface equations are numbered after the non-interface equations are reordered.

The active column matrix storage is used for the substructure stiffness matrices. In the active column storage scheme as shown in Figure 3.1, only the elements up to the first non-zero element in each column are stored in a vector form. The location of the diagonal elements is stored in another vector (ColId).

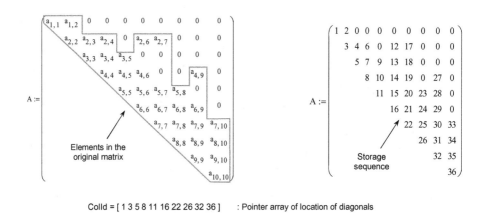

ColId = [1 3 5 8 11 16 22 26 32 36]   : Pointer array of location of diagonals

A = [$a_{1,1}$ $a_{1,2}$ $a_{2,2}$ $a_{2,3}$ $a_{3,3}$ ...]   : Elements of the original matrix stored in vector format

Figure 3.1 Active Column Storage

The LU decomposition of the matrix A is actually the result of Gaussian elimination process where the unit diagonal lower triangular matrix L consists of the row multipliers and the upper triangular matrix U is the resultant matrix after eliminations.

The internal degrees of freedom are condensed by using Equations (3.1) to (3.5) shown below. The first step of condensation is the factorization where the upper and lower triangular parts of the stiffness matrix are calculated column by column by using Equations (3.1) and (3.2). During the update of the upper triangular coefficients belonging to the interface degrees of freedom, only the coefficients that describe the coupling between the internal and interface degrees of freedom are considered.

$$U_{ij} = A_{ij} - \sum_{k=kf}^{kl} L_{ki} \cdot U_{kj} \qquad j=2 \text{ to } NEQ \qquad i=jz \text{ to } j \qquad (3.1)$$

$$L_{ji} = U_{ij}/D_{ii} \qquad\qquad i=iz \text{ to } LEQ \qquad j=jz \text{ to } i\text{-}1 \qquad (3.2)$$

where *NEQ* = *total number of equations*
LEQ = *number of equations to be reduced (number of internal dofs)*
A = *Symmetric stiffness matrix, active column storage*
U = *upper diagonal part*
L = *lower diagonal part*
iz = *location of first non-zero column i*
jz = *location of first non-zero column j*
kf = *the maximum of iz or jz*
kl = *the minimum of either 1 or LEQ*

The factorization is followed by forward substitution, or load factorization where the internal loadings are transferred to the interfaces by using Equations (3.3) and (3.4):

$$F_i = F_i - \sum_{k=jz}^{kff} L_{ki} \cdot F_k \qquad i=1 \text{ to } NEQ \qquad (3.3)$$

$$F_i = F_i / D_{ii} \qquad i=1 \text{ to } LEQ \qquad (3.4)$$

where F = load vector

kff = the minimum of jz or the height of the column j starting from LEQ

After obtaining the displacements for the interface degrees of freedom, the back substitution, or recovery, is performed where the displacements for the internal degrees of freedom are calculated by using Equation (3.5):

$$Fu_k = Fu_k - L_{ki} \cdot Fu_i \qquad i=NEQ \text{ to } 1 \qquad k=iz \text{ to } kff \qquad (3.5)$$

where Fu = displacement vector

3.2.2 Operation Count

Operation count calculations indicate how different variables affect the solution time. However, developing an analytical method to estimate the number of operations for the above algorithm is difficult due to the complexity of the algorithm. First of all, the active column storage scheme has variable column heights which may vary considerably for a given problem and also from problem to problem. Due to the variation of column heights, the average column height is used throughout the calculations below. Furthermore, the current condensation algorithm requires the interface degrees of freedom to be stored at the end of the matrix. During the equation numbering, the internal equations are numbered by using a profile minimization algorithm and then the interface equations are numbered. These independent numberings usually cause a significant increase of the

78

average column heights of the interface equations and the resulting stiffness matrix looks like an arrow as shown in Figure 3.2.

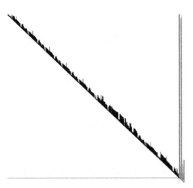

Figure 3.2 An Example Profile of a Stiffness Matrix Assembled for Condensation

In light of the difficulties noted above, the operation count calculations in this study are performed by using four different variables: the average column height of the internal and interface degrees of freedom, and the number of the internal and the interface degrees of freedom. These variables are illustrated in Figure 3.3:

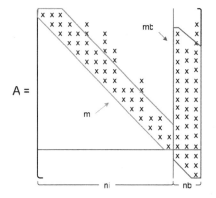

mi = *average column height of internal dofs*

mb= *average column height of interface dofs*

ni = *number of internal dofs*

nb = *number of interface dofs*

Figure 3.3 A Typical Shape of Assembled Substructure Stiffness Matrix

The entire condensation algorithm is examined in three steps for the operation count calculations, i.e., factorization, forward substitution (load factorization), and back substitution (Recovery).

The factorization algorithm is examined in two parts. The first part involves only the internal equations, i.e. equations from 1 to 'ni', whose average column heights are constant and equal to 'mi'. The pseudo code that calculates the upper and lower triangular coefficients is presented in Figure 3.4. The stiffness matrix is stored in one-dimensional array called 'M' and the location of the diagonals is kept in an array called 'loc'.

```
for i=2 to ni {
  jj=i-mi+1
  if (jj<1) jj=1

  // upper triangular part
  for j=jj to i-1 {
      nDot=mi-i+j-1  // number of data in dot product

      // calculate the addresses of the columns i and j
      nStartJ=loc[j]-nDot
      nEndJ  =loc[j]-1
      locIJ  =loc[i]-i+j    // location of jth element of ith column
      nStartI=locIJ-nDot
      // dot product
      S=0
      for k=1 to nDot
         S=S + M[nStartI+k] * M[nStartJ+k]

      M[locIJ]=M[locIJ]-S
  }

  // lower triangular part
  S=0
  for n=nStartJ to nEndJ {
      locDiag=loc[jj]
      jj=jj+1
      T=M[n]
      M[n]=M[n]/M[locDiag]
      S=S+M[n]*T
  }

  M[locIJ]=M[locIJ]-S
}
```

Figure 3.4 Pseudo Code for Factorization up to column 'ni'

In this part of the algorithm, columns from 2 to 'ni' are factorized. For the ith column, the algorithm first calculates the upper triangular part, then calculates the lower triangular part, and finally updates the column's diagonal. In order to update the elements of the ith column, the lower triangular coefficients of the columns from '(i-mi)' to '(i-1)' are used. For that case, the number of elements that will be involved in the dot product will be equal to 'mi-i+j'. The dot product operation will cost one addition and one multiplication per element. Therefore, the total number of operations for each dot product will be equal to '2.(mi-i+j)' elements. The result of the dot product of the column i and column j, is used to update the jth element of the ith column. Thus, after the dot product, the result is subtracted from it. The total number of operations for the update of the upper triangular elements of the internal equations (T_{u1}) is calculated as:

$$T_{u1} = \sum_{i=2}^{ni} \sum_{j=i-mi}^{i-1} 2 \cdot (mi - i + j) + ni \tag{3.6}$$

$$T_{u1} = mi^2 \cdot (ni - 1) - mi \cdot (ni - 1) + ni \tag{3.7}$$

The lower triangular coefficients of the off-diagonals of the ith column are calculated by dividing its elements with the corresponding diagonals. At the same time, the coefficient that will modify the ith diagonal is found by multiplying the 'L' of the ith column with the 'U' of it. Thus, there will be one division to calculate each element of 'L', one multiplication and one addition to calculate its effect on the coefficients. As a final step, the ith diagonal is updated by subtracting this coefficient from it. Thus, a total of three operations will be performed for each element of the ith column. The number of operations for calculating the lower triangular coefficients of the internal equations (T_{11}) is given in Equation (3.8):

$$T_{l1} = ni + \sum_{j=1}^{ni} 3 \cdot mi \qquad (3.8)$$

$$T_{l1} = 3 \cdot mi \cdot ni + ni \qquad (3.9)$$

The second part of the factorization algorithm, factorizes the equations at the interfaces. The operation count calculations for this part are more complicated due to the two different average column heights, 'mi' and 'mb'. Moreover, all the dot products and the lower triangular coefficient calculations use the elements above the row 'ni'. For that reason, the upper triangular coefficient calculations are examined in three regions. While defining these regions, it is assumed that the average column height at the interface is larger than the number of interface degrees of freedom. The boundaries of these regions are defined as:

1. $j < ni$ and mb-i-j $< mi$

2. $j < ni$ and mi $<$ mb-i-j

3. $j > nb$

The pseudo code for the second part of the factorization algorithm is given in Figure 3.5. It is very similar to the code for the first part except the number of elements involved in the dot product changes from region to region.

For each region, the number of elements in dot products is equal to 'mb-i-j-1', 'mi-1', 'mb+ni-i+1', respectively. The calculations are the same as the first part but this time

83

only the columns from 'ni+1' to 'n' are factorized. The total number of operations to update the upper triangular part of the interface equations (T_{u2}) is given in Equation (3.10).

$$T_{u2} = \sum_{i=ni+1}^{ni+nb} \left[\sum_{j=i-mb+1}^{mi+i-mb} 2 \cdot (mb-i+j-1) + \sum_{j=-mb+mi+i+1}^{ni} 2 \cdot (mi-1) + \sum_{j=ni+1}^{i-1} 2 \cdot (mb+ni-i-1) \right] + nb \qquad (3.10)$$

$$T_{u2} = 2 \cdot mi \cdot mb \cdot nb + (mb-mi) \cdot nb^2 - (mi^2 + 3 \cdot mb) \cdot nb - \frac{2}{3} \cdot nb^3 + \frac{11}{3} \cdot nb \qquad (3.11)$$

Only the lower triangular coefficients of the elements above the row 'ni' are calculated for the ith column. Likewise, the same calculations are performed as in the first part. The number of operations to calculate the lower triangular coefficients of the interface equations (T_{l2}) is equal to:

$$T_{l2} = nb + \sum_{ni+1}^{ni+nb} \sum_{i-mb}^{i-1} 3 \qquad (3.12)$$

$$T_{l2} = 3 \cdot mb \cdot nb + nb \qquad (3.13)$$

```
for i=ni+1 to n {
  jj=i-mb+1
  if (jj<1) jj=1

  // upper triangular part
  for j=jj to i-1 {
      if (j ≤ ni) {
        if (j ≤ i-mb+mi)
           nDot=mb-i+j-1  // region 1
        else
           nDot=mi-1       // region 2
      } else {
           nDot=mb+ni-i+1    // region 3
      }
      // calculate the addresses of the columns i and j
      nStartJ=loc[j]-nDot
      nEndJ  =loc[j]-1
      locIJ  =loc[i]-i+j
      nStartI=locIJ-nDot
      // dot product
      S=0
      for k=1 to nDot
         S=S + M[nStartI+k] * M[nStartJ+k]
      M[locIJ]=M[locIJ]-S
  }
  // lower triangular part
  S=0
  if (j>ni) nEndJ=nEndJ+ni-j+1
  for n=nStartJ to nEndJ {
    locDiag=loc[jj]
      jj=jj+1
      T=M[n]
      M[n]=M[n]/M[locDiag]
      S=S+M[n]*T
  }
  M[locIJ]=M[locIJ]-S
}
```

Figure 3.5 Pseudo Code for Factorization for Equations Between 'ni' and 'n'

Finally, the total number of operations (T_{fact}) for the factorization is found by summing Equations (3.6), (3.8), (3.10) and (3.12). The result is:

$$T_{fact} = T_{u1} + T_{l1} + T_{u2} + T_{l2} \tag{3.14}$$

$$T_{fact} = mi^2 \cdot (ni - nb - 1) + 2 \cdot mi \cdot mb \cdot nb + (mb - mi) \cdot nb^2 +$$
$$2 \cdot ni \cdot (mi + 1) - \frac{2}{3} \cdot nb^3 + \frac{14}{3} \cdot nb \tag{3.15}$$

The next step is the factorization of the load vectors. The pseudo code shown below in Figure 3.6 presents the forward substitution algorithm for a single loading condition. The same algorithm is used for all other loading conditions. The loadings are stored in a vector called 'R'.

The forward substitution algorithm is examined in two parts also. The first part uses the factorized equation up to 'ni' to modify the loads. The second part uses the interface equations. This time, only the element above row 'ni' is considered in dot products. The final step of the forward substitution is the division of the internal forces with the corresponding diagonal elements. The total number of operations for the forward substitution (T_{fs}) is calculated in Equation (3.17):

$$T_{fs} = b \cdot \left[\sum_{2}^{ni} 2 \cdot mi + \sum_{i=ni+1}^{ni+nb} 2 \cdot (mb + ni - i) + \sum_{2}^{ni+nb} 1 + \sum_{1}^{ni} 1 \right] \tag{3.16}$$

$$T_{fs} = b \cdot \left(2 \cdot mi \cdot ni + 2 \cdot mb \cdot nb + 2 \cdot ni - 2 \cdot mi - nb^2 - 1 \right) \tag{3.17}$$

86

```
// forward substitution for a single load vector
// first part
for i=2 to ni {

  jj=i-mi+1
  if (jj<1) jj=1

  nStart=loc[i]-mi

  S=0
  for k=jj to i
      S=S + M[nStart+k] * R[k]

  R[i]=M[i]-S
}
// second part
for i=ni+1 to n {

  jj=i-mb+1
  if (jj<1) jj=1

  nStart=loc[i]-mb
  locEnd=mb+ni-i+1

  S=0
  for k=jj to locEnd
      S=S + M[nStart+k] * R[k]

  R[i]=M[i]-S
}

for i=1 to ni
  loc=Loc[i]
  R[i]=R[i]/M[loc]
}
```

Figure 3.6 Pseudo Code for Forward Substitution

The recovery or the back substitution is performed after the interface displacements have been calculated. Both recovery and forward substitution algorithms perform nearly the same operations. The only difference is that the forces are not divided by the diagonals during recovery. The number of operations for back substitution (T_{bs}) is calculated by using Equation (3.18):

$$T_{bs} = b \cdot \left[\sum_{2}^{ni} 2 \cdot mi + \sum_{i=ni+1}^{ni+nb} 2 \cdot (mb + ni - i) \right] \qquad (3.18)$$

$$T_{bs} = b \cdot \left(2 \cdot mi \cdot ni + 2 \cdot mb \cdot nb - 2 \cdot mi - nb^2 - nb \right) \qquad (3.19)$$

Operation count calculations show that the column heights of interface equations play a significant role on the condensation time. The number of operations during factorization has an order of either $O(mi^2(ni-nb))$ or $O(2mi.mb.nb)$ or $O((mb-mi)nb^2)$. The factorization time can be governed by any combination of these products. If the number of interface equations is kept constant, there will be a linear relationship between the solution time and the average column height at the interface. Similarly, for a constant number of equations, having more interface equations increases the factorization time because of higher column heights of the interface equations.

These equations also show why the current partitioning algorithms can not successfully balance the computational workloads for direct condensation. First of all, equating the number of nodes or elements in each substructure does not mean that their internal and interface column heights will be the same or even close to the same value. Moreover, there is a non-linear relationship between the factorization time and the

88

average internal column heights and the number of interface equations. Small differences in these values can create large imbalances. The partitioning algorithms can produce substructures having similar average internal column heights for uniform structures. Nevertheless, this time, the number of interface equations of substructures differs as the number of partitions increase. Finally, the average column heights of the substructures can only be calculated after partitioning and their effects on the factorization time can not be ignored.

3.3 Condensation Time Estimations

3.3.1 Time Estimations for Uniform Matrices

The operation count equations analytically indicate the effects of different variables on the condensation time. During the derivation of these equations, it was assumed that the processor performs all the computations at a constant speed. However, with all the variations in processor types, bus speeds, memory speeds, and cache memory sizes, this may not be valid for different PC configurations.

Thus, in order to see whether the constant processor speed assumption is valid for this algorithm and also to verify the correctness of the above equations, several test runs were performed on various machines. Then, the actual condensation times were compared with the analytical condensation times estimated using the operation count equations.

As a first step, various stiffness matrices of different number of degrees of freedom with constant internal and interface column heights were generated. Then, the factorization speed was calculated by averaging the speed values obtained by dividing the corresponding operation count values using Equations (3.7) and (3.11) with the time spent to calculate the upper triangular elements from Equation (3.1) for each matrix. Next, a series of condensations were performed on various stiffness matrices by changing the interface column height while keeping the number of internal and interface equations, and the internal column height constant. As a final step, the same procedure was repeated by changing the number of interface equations only.

First, runs were executed using on one of the computers in an old cluster (DEC) that had a Pentium 166 processor with 64 Mb RAM. The computer operating system was Windows NT and the external bus speed was 66 Mhz. Figure 3.7 shows the results obtained for both series of runs. The '+' signs indicate the actual factorization times (experimental) whereas the solid line is the estimated time (analytical) which was calculated by dividing the operation count value with the average factorization speed. A rather small matrix size was chosen for this computer due to its limited memory.

The graph at the top in Figure 3.7 was obtained by changing the interface column height from 500 to 2500. The factorization speed was found as 16.66 Mflops (floating point operations per second) and used for both graphs. The estimated factorization times were larger than the actual times for column heights up to 1200, but after that point, the time estimations were smaller. This showed that the factorization speed dropped as the size of the problem increased for this computer. The maximum error in the estimations was 13.7% and the average error was 5.3%. The graph at the bottom shows the analytical estimations and actual factorization times as the number of interface equations varied from 50 to 1000 for a constant interface column height of 1000. For this case, the analytical timing results overestimated the actual times within an average error of 6.3%.

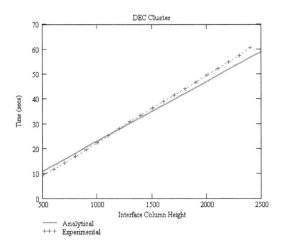

Factorization times with varying interface column height

Factorization speed (Mflops)	16.66		
Number of internal equations	5000	Internal Column Height	150
Number of interface equations	500	Interface Column Height	500-2500

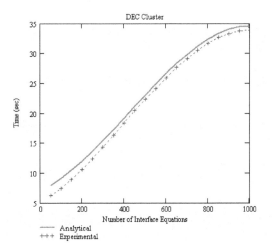

Factorization times with varying number of interface equations

Factorization speed (Mflops)	16.66		
Number of internal equations	5000	Internal Column Height	150
Number of interface equations	50-1000	Interface Column Height	1000

Figure 3.7 Factorization Times on 166 Mhz DEC Computer

91

Similar runs were performed on one of the computers of another cluster (AFC). This computer had a 400 Intel Celeron processor with 128 MB RAM and 128KB L2 Cache. The external bus speed was 66 Mhz and the operation system was Windows 2000. The experimental and the analytical results are presented in Figure 3.8.

The factorization speed for this computer was computed as 38.85 Mflops which was approximately 2.4 times faster than the DEC computer. The time estimations were much more satisfactory for these machines. The average error was 1% as the interface column height varied and 3% as the number of interface equations varied.

The third series of runs were performed on a computer (DELL) which had a 2.4GHz Intel Pentium 4 processor with 1GB Ram, 512 KB L2 cache and 800 Mhz bus speed. The operating system on this machine was Windows XP Professional Edition.

The factorization time results with varying interface column height and number of interface equations were given in Figure 3.9. The factorization speed was computed as 166.58 Mflops which was 10 times faster than the 166 Mhz DEC computer and 4.3 times faster than the 400 Mhz Celeron AFC computer. Similar to the AFC results, the estimations were in a good agreement with the actual condensation times. The average errors were 0.5% and 0.1% for the graphs at the top and bottom, respectively.

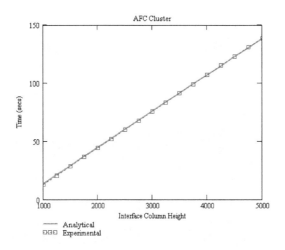

Factorization times with varying interface column height

Factorization speed (Mflops)	38.85		
Number of internal equations	10000	Internal Column Height	100
Number of interface equations	1000	Interface Column Height	1000-5000

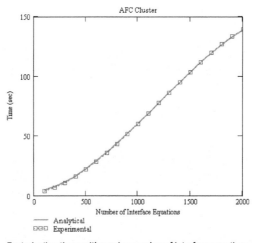

Factorization times with varying number of interface equations

Factorization speed (Mflops)	38.85		
Number of internal equations	10000	Internal Column Height	100
Number of interface equations	100-2000	Interface Column Height	2500

Figure 3.8 Factorization Times on 400 MHz Celeron Computer (AFC)

93

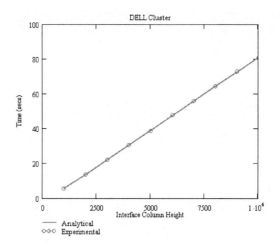

Factorization times with varying interface column height

Factorization speed (Mflops)	166.58		
Number of internal equations	10000	Internal Column Height	200
Number of interface equations	1000	Interface Column Height	1000-10000

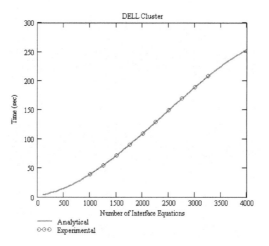

Factorization times with varying number of interface equations

Factorization speed (Mflops)	166.58		
Number of internal equations	10000	Internal Column Height	200
Number of interface equations	100-4000	Interface Column Height	5000

Figure 3.9 Factorization Times on 2.4 GHz Pentium 4 Computer (DELL)

94

In order to test the forward and back substitution performances of these three computers, similar test runs were performed. This time, a constant size stiffness matrix was condensed with varying number of loading conditions. For each computer, the forward and back substitution speeds were computed separately. This is mainly because the governing operation for forward substitution is a dot product, Equation (3.3), whereas back substitution requires the summation of a vector-scalar product Equation (3.5).

Figures from 3.10 to 3.12 present the results on DEC, AFC and DELL computers, respectively. The graph at the top shows the forward substitution times and the graphs at the bottom show the back substitution time.

The forward and back substitution speeds of DEC and AFC computers were close, but the DELL computer performed back substitutions 3 times faster than forward substitution. This indicates that the DELL computer performs scalar vector multiplications much faster than vector-vector multiplications. The factorization and the forward substitution speeds for all computers were very close. The maximum difference encountered was for the DEC computer because its speed varied as the size of the matrix increased. Moreover, the time estimations for both forward and back substitution were in a good agreement with the actual times for all computers.

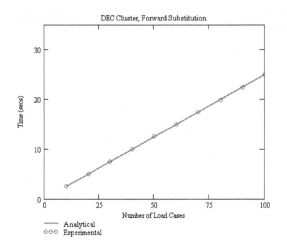

Forward substitution times for different number of load cases

Forward substitution speed (Mflops)	15.05	Number of Load Cases	10-100
Number of internal equations	5000	Number of interface equations	500
Internal Column Height	150	Interface Column Height	2500

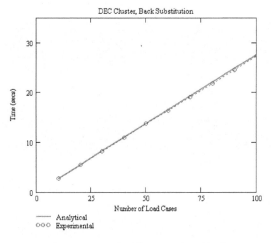

Back substitution times for different number of load cases

Back substitution speed (Mflops)	13.56	Number of Load Cases	10-100
Number of internal equations	5000	Number of interface equations	500
Internal Column Height	150	Interface Column Height	2500

Figure 3.10 Forward and Back Substitution Times on 166 Mhz Computer (DEC)

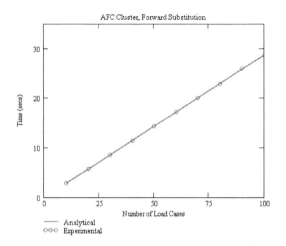

Forward substitution times for different number of load cases

Forward substitution speed (Mflops)	38.39	Number of Load Cases	10-100
Number of internal equations	10000	Number of interface equations	100-2000
Internal Column Height	100	Interface Column Height	2500

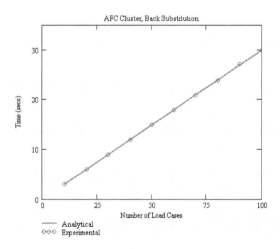

Back substitution times for different number of load cases

Back substitution speed (Mflops)	36.72	Number of Load Cases	10-100
Number of internal equations	10000	Number of interface equations	100-2000
Internal Column Height	100	Interface Column Height	2500

Figure 3.11 Forward and Back Substitution Times on 400 Mhz Computer (AFC)

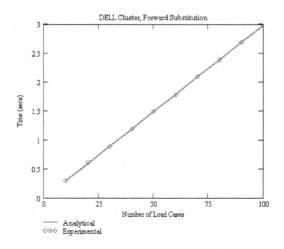

Forward substitution times for different number of load cases			
Forward substitution speed (Mflops)	168.04	Number of Load Cases	10-100
Number of internal equations	10000	Number of interface equations	100-4000
Internal Column Height	200	Interface Column Height	5000

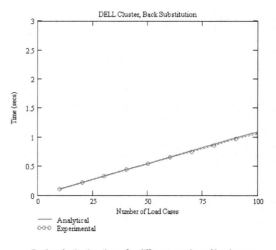

Back substitution times for different number of load cases			
Back substitution speed (Mflops)	462.75	Number of Load Cases	10-100
Number of internal equations	10000	Number of interface equations	100-4000
Internal Column Height	200	Interface Column Height	5000

Figure 3.12 Forward and Back Substitution Times on 2.4 GHz Computer (DELL)

These runs showed that the factorization, forward and back substitution times are predictable within 1% error range. The computation speeds are constant, in other words, the speed does not depend on the density or the size of the matrix. Thus, once the computation speed for an operation is found, it is possible to obtain good time estimations for the condensation. On the other hand, the computation speeds can vary for different operations, such as factorization and back substitution, depending on the computer hardware. Therefore, the computation speed of each operation must be computed separately.

3.3.2 Time Estimation for Non-uniform Matrices

Up to this point, matrices having constant column heights were utilized in the calculations; however, real problems have variable column heights. Nikishkov et al. [72] introduced a matrix density constant in the operation count calculations to consider such variations. Their approach is still approximate and it is possible to obtain an exact operation count numerically for such matrices.

In order to obtain the operation count for factorization, only the computation of the upper triangular part of the stiffness matrix was considered, Equation (3.1), since for large size problems, the time spent to calculate the lower triangular elements becomes insignificant. For example, the number of operations to compute the upper triangular elements of the internal equations (Equation (3.7)) is 'mi/4' times larger than the number of operations required to compute their lower triangular coefficients (Equation (3.9)).

The operation count is computed numerically as follows. First, the internal i^{th} column is factorized by the previous columns starting from the jth column. The jth column is found by subtracting the column height of the ith column from 'i'. Each element of the ith column is updated by subtracting the result of the dot product The number of elements in the dot product is found by Equation (3.20) for updating the jth element of column 'i':

$$nd_i = min(m_i - i + j, m_j)$$ (3.20)

The factorization of interface equations is similar. This time, if the jth column is at the interface, the number of elements in the dot product is calculated with Equation (3.21):

$$nd_i' = nd_i + ni - j$$ (3.21)

The number of operations required to update the jth element of the ith column is equal to two times the 'nd' value. Thus, the total operation count for factorization is equal to the summation of the number of operations to update every element of the stiffness matrix which is calculated using Equation (3.22):

$$OC = \sum_{i=1}^{ni} \sum_{j=i-m_i}^{i} 2 \cdot nd_j + \sum_{i=ni+1}^{n} \sum_{j=i-m_i}^{i} 2 \cdot nd_j'$$ (3.22)

The governing operation is the dot product operations for forward substitution, Equation (3.3), and the summation of scalar-vector products, Equation (3.5), for back substitution. For both cases, the number of elements in the governing operations is the same for the ith column and found by Equation (3.23) if 'i' is an internal equation. If 'i' is an interface equation, Equation (3.24) will be utilized.

$$nd_i = m_i \tag{3.23}$$

$$nd_i' = nd_i + ni - i \tag{3.24}$$

Thus, the total number of operations during forward and back substitution for a single loading condition is computed by Equation (3.25):

$$OC = \sum_i^{ni} 2 \cdot nd_i + \sum_{i=ni+1}^{n} 2 \cdot nd_i' \tag{3.25}$$

In order to determine if the time estimation approach is valid for actual problems, a 2D square membrane was solved by dividing the mesh into 4 substructures as shown in Figure 3.13. The structure contained 26,266 nodes with two degrees of freedom per node resulting in a total number of equations of 52,526.

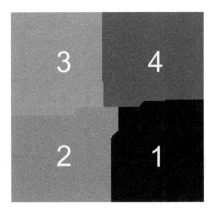

Figure 3.13 Membrane Problem, 2D 160x160 Mesh with 4 substructures

Tables from 3.1 to 3.3 show the estimated and actual condensation times of four substructures solved on a single computer belonging to DEC, AFC, and DELL

101

computers, respectively. The operation count values were computed numerically. The time estimations was found by dividing the operation count values with the factorization speeds computed in Section 3.4

Table 3.1 Factorization Times, DEC Cluster

Substructure	Operation Count	Estimated Factorization Time	Actual Factorization Time	Error
1	7.898177×10^8	47.41 secs	53.797 secs	-13.5%
2	8.319888×10^8	49.94 secs	54.218 secs	-7.9%
3	7.843286×10^8	47.08 secs	51.574 secs	-8.7%
4	8.038269×10^8	48.25 secs	52.735 secs	-8.3%

Table 3.2 Factorization Times, AFC Cluster

Substructure	Operation Count	Estimated Factorization Time	Actual Factorization Time	Error
1	7.898177×10^8	20.33 secs	18.016 secs	+11.4%
2	8.319888×10^8	21.41 secs	19.168 secs	+10.5%
3	7.843286×10^8	20.19 secs	18.116 secs	+10.3%
4	8.038269×10^8	20.69 secs	20.219 secs	+2.3%

Table 3.3 Factorization Times, DELL Cluster

Substructure	Operation Count	Estimated Factorization Time	Actual Factorization Time	Error
1	7.898177×10^8	4.74 secs	4.81 secs	-1.4%
2	8.319888×10^8	4.99 secs	5.06 secs	-1.4%
3	7.843286×10^8	4.71 secs	4.78 secs	-1.5%
4	8.038269×10^8	4.82 secs	4.87 secs	-1.0%

The actual factorization times were slower than the estimated ones in DEC computer whereas the results were the opposite for the AFC cluster. For both cases, the error is approximately 10%. The results in the DELL computer were much better. The maximum deviation from the actual times was 1.5%.

Table 3.4 presents the times for forward substitution on AFC computer. This time, the 2D membrane problem was solved for 50 load cases. The operation count values were computed using Equation (3.25), and divided by the forward substitution speed In this case, the time estimations are smaller than the actual times but the difference is still approximately 10%.

Table 3.4 Forward Substitution Times with 50 load cases, AFC Cluster

Substructure	Operation Count	Estimated Forward Substitution Time	Actual Forward Substitution Time	Error
1	3.13277×10^8	8.16 secs	9.07 secs	-10.0%
2	3.17099×10^8	8.26 secs	9.15 secs	-9.7%
3	3.11795×10^8	8.12 secs	8.99 secs	-9.7%
4	3.13992×10^8	8.18 secs	9.04 secs	-9.5%

The following two tables, Tables 3.5 and 3.6, show the results for forward and back substitutions, respectively, on DELL computer. The operation count values were divided with the forward and back substitution speeds in order to compute the substitution time estimations. The estimated times for both forward and back substitutions are very close to the actual times. The comparisons indicate that all results were within 5%. Moreover, the DELL computer performed the back substitutions three times faster than the forward substitutions.

Table 3.5 Forward Substitution Times with 50 load cases, DELL Cluster

Substructure	Operation Count	Estimated Forward Substitution Time	Actual Forward Substitution Time	Error
1	3.13277×10^8	1.85 secs	1.89 secs	-2.1%
2	3.17099×10^8	1.87 secs	1.91 secs	-2.1%
3	3.11795×10^8	1.84 secs	1.89 secs	-2.6%
4	3.13992×10^8	1.86 secs	1.89 secs	-1.6%

Table 3.6 Back Substitution Times with 50 load cases, DELL Cluster

Substructure	Operation Count	Estimated Back Substitution Time	Actual Back Substitution Time	Error
1	3.13277×10^8	0.68 secs	0.66 secs	+2.9%
2	3.17099×10^8	0.68 secs	0.66 secs	+2.9%
3	3.11795×10^8	0.67 secs	0.64 secs	+4.7%
4	3.13992×10^8	0.68 secs	0.64 secs	+5.9%

3.4 Out-of-Core Condensation

A major disadvantage of the direct solution methods is their need of large free in-core memory. The in-core memory is actually the real physical memory (RAM) without the consideration of the virtual memory. Although the structure is divided into smaller substructures during a substructure based solution approach, the available in-core memory may not be enough for the solution of large problems using a low number of processors. For such cases, an out-of-core version of the presented algorithm was utilized.

In the out-of-core version, the stiffness matrix is stored on a disk. It is divided into predetermined size parts, called blocks. During the factorization, each block is loaded into memory as needed. Once the blocks are factorized, they are written back to the disk and the new block is loaded into memory.

The factorization starts by loading the first block into the memory and calculating its upper triangular coefficients. After that, the blocks to be updated by the columns of the first block are loaded into the memory and updated one-by-one. Once all blocks have been updated, the next block is loaded into memory for factorization. This process continues until all blocks are factorized.

The following Table 3.7 shows the out-of-core factorization times of the high-rise building I model, Appendix A.1.4, with 4 processors on DELL cluster. The fourth column shows the total number of read and write operations performed during the factorization. The speed values are calculated by dividing the operation count with the condensation times. Each block was approximately 64 Mbytes.

Table 3.7 Out-of-Core Factorization

Processor	Op.Count	# of Blocks	# of I/O	Time (seconds)	Speed (Mflops)
1	7.48×10^{11}	37	533	5605.1	133.45
2	2.93×10^{11}	23	156	1956.9	149.72
3	3.29×10^{11}	25	216	2329.6	141.22
4	6.39×10^{11}	35	509	5170.5	123.70

The second and the third factorizations were the fastest ones due to fewer blocks; thus, the time spent during the block read/write was smaller. The factorization speeds decreased as the number of blocks increased. The first and the fourth substructures had more blocks; thus, their factorization speeds were slower. Although the fourth substructure required less I/O than the first substructure during factorization, its factorization speed was slower. Still, it is possible to say that the out-of-core factorization time not only depends on the operation counts but also the number of blocks and the total number of read and write operations. Thus, the factorization speed varies depending on the size and the density of the matrices for out-of-core condensation.

3.5 Conclusions

These test runs showed that it was possible to predict the factorization, forward and back substitution times of matrices stored with the active column storage scheme. The factorization, forward and back substitution speeds are mostly constant as the properties of the matrix change, but they don't have to be the same. For example, although the factorization, forward and back substitution speeds of both DEC and AFC computers were close, the DELL computer performed back substitutions three times faster than the factorization and the forward substitution. As shown in the previous sections, even though the processor speed of the DELL computer was 6 times faster than the AFC computer, the DELL computer performed factorizations only 4.3 times faster. On the other hand, its back substitution speed was 13 times faster than AFC computer. Hence, test runs are the safest way to investigate the computational speed of a computer. They should be performed for the factorization, forward and back substitution, with various matrix sizes. Once the average speeds are obtained, the solution times can be predicted within an acceptable error range (<15%).

The interface equations had a significant effect on the factorization time. Their column heights were much larger than the column heights of the internal equations; thus, they required more operations during factorizations. They are also the major sources of imbalances in the condensation times of substructures created with a partitioning algorithm because many partitioning algorithms equate the number of nodes or elements in the substructures but not the number of interface nodes. Even if a structure is partitioned in such a way that each substructure has an equal number of interface nodes, their average interface column heights may be substantially different. As a result, any partitioning algorithm that attempts to create balanced substructures for direct condensation should consider the number of interface nodes and also the average column heights of the interface equations.

CHAPTER 4

WORKLOAD BALANCING FOR DIRECT CONDENSATION

4.1 Introduction

In the substructure based solution approach, the structure is first divided into substructures. Then, the substructure level stiffness matrices and force vectors are assembled and the internal equations are transferred to the substructure interfaces using condensation algorithms. As a result, the solution is converted into the solution of dense interface equations. After having computed the interface displacements, each substructure's internal displacements are computed. In the parallel implementation of this approach, the substructures are distributed to processors and both the condensations and the solution of interface equations are performed simultaneously.

The direct solvers are often used to condense the substructures' stiffness matrices at substructure based solution approaches [73] since they are robust and very suitable for problems having multiple loading conditions. On the other hand, any imbalance in the condensation times of the substructures reduces the efficiency of the parallel solution since the interface solution can not begin until all processors the finish condensation. Hence, the partitioning of the structure into substructures is crucial for such methods.

108

In the literature, various partitioning approaches exist [17]. The basic goal of many of the partitioning approaches is to minimize the communication between processors while keeping a balanced number of elements or nodes in each partition. This goal can be achieved if the computational cost can be represented by a single weight value assigned to a node or an element. However, when a direct condensation method is used, such weight definitions are insufficient to provide a balanced distribution of the computational load [19]. There are other variables that affect the condensation time such as equation numbering, the profile of the stiffness matrix, and the number of the internal and interface equations. Moreover, the variables that affect the condensation time depend on the way in which the structure is partitioned. In other words, the computational load of each substructure can only be estimated after partitioning. Secondly, the profile of the stiffness matrix depends on how the equations are numbered. If an active column storage scheme is used, the equation numbering is generally performed by using a type of profile minimization algorithm. Such algorithms are based on heuristic approaches. Therefore, it is very difficult to predict the effect of any partition changes on the equation numbering and hence on the condensation time. Thus, it is rather complicated to partition a structure while balancing the workload for direct condensation.

In order to overcome such difficulties, Fulton and Su [82], developed a dynamic processor assignment strategy. Once the structure is partitioned, they estimated the computational loads for condensing each substructure's stiffness matrix and assigned more processors to the ones which were estimated to require more computation. The algorithm was designed to work on shared memory computers. Yang and Hsieh [102] proposed an iterative partitioning approach that attempted to balance the condensation times of substructures by modifying the partitions considering their elements' weights. At each iteration, each substructure's workload was estimated by computing the operation count for sparse matrix condensation and the weight of the elements within a substructure was modified according to each substructure's workload. Then, either the structure was repartitioned or the substructures were modified by element migration. The iterations continued until a desired balance was obtained. For the test problems considered by Yang and Hsieh [102], this approach provided more balanced substructures and decreased the condensation time. On the other hand, the time consumed during the iterations was so high, the method was suitable only for nonlinear dynamic analysis and not for a linear static analysis.

This chapter focuses on the development of a fast and efficient workload balancing method for direct condensation that can used for solving large linear systems on PC clusters. The method iteratively searches for more balanced substructure workloads by modifying them according to their estimated condensation time ratios and is similar to the work by Hsieh et al. [98]. However, the method in this work is designed for the active column matrices and utilizes nodal graphs in order to work with models composed of both frame and shell elements as discussed in Section 2.2.1. Moreover, the framework has a parallel structure. In other words, all the computations during the iterations are performed in parallel. This way, the time consumed during the workload balancing step is decreased and the algorithm becomes more suitable for large linear static problems.

4.2 Method

The workload balancing framework was designed as an independent program that prepares input data for the parallel substructure based solver. The MPICH [115] message passing library was used for parallelism. The program was developed with C++ and FORTRAN programming languages and implemented under the Windows operating system.

The next section gives a brief description about the workload balancing algorithm with the following sections providing a detailed description of each step.

4.2.1 Workload Balancing Algorithm

The flow chart of the algorithm is presented in Figure 4.1. The solution begins by converting the structural information into the nodal graph representation. Then, the nodal graph is partitioned into 'n' parts by using METIS [30] where 'n' is equal to the number of available processors. After assigning a single substructure to each processor, the nodal graph and the initial partitioning information are distributed to every processor. The processors first extract their assigned substructure's subgraphs from the nodal graph using the partitioning information. A subgraph is actually the nodal graph of a substructure with links to adjacent substructures. Then, each processor optimizes the equation numbering of their substructure and calculates the operation counts for condensation. During both computations, it is assumed that there is a single degree of freedom per node.

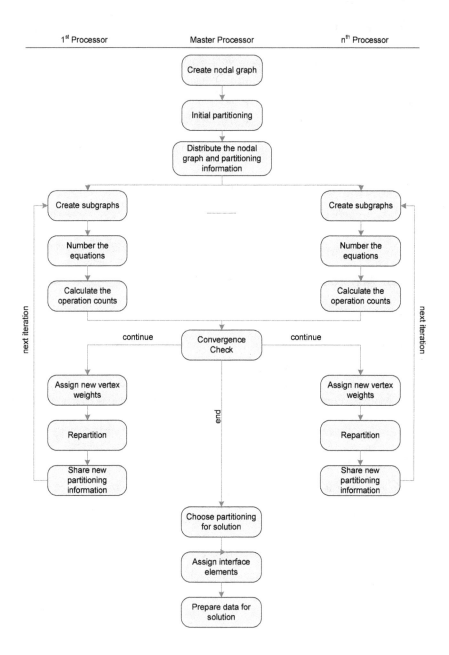

Figure 4.1 The Flow Chart of the Workload Balancing Algorithm

Next, the master processor, which performs the initial partitioning, collects the operation count values and checks whether the workload is balanced or the maximum number of iterations has been reached. If so, the iterations are finalized and the structural data for the solution is prepared. Otherwise, the master processor calculates the imbalance factor for each substructure and distributes it to other processors. Each processor then calculates new vertex weights of their substructures by using the imbalance factor and updates the weights of their vertices. Then, the current partitioning information is stored and repartitioning is initiated. The substructures are repartitioned by using either diffusion or scratch-remap type repartitioning algorithms of the PARMETIS [28] library. Once the new partitioning information is obtained, it is distributed to all other computers and the next iteration starts.

When the iterations are finalized, the master processor scans all partitioning results created during the iterations and chooses the one that provides the best estimate of the solution time. The solution time estimation can consider the estimation of either the condensation time only or the total solution time including the interface solution time.

The final step is the preparation of the structural data where the subgraphs are converted into nodes and element definitions. During that process, the interface elements, whose nodes are on two or more substructures, are assigned to one of their adjacent substructures according to two criteria which will be discussed later.

4.2.2 Data Structure

The first step of the algorithm is the conversion of the structural information, i.e. element connectivity, into a nodal graph representation. In a nodal graph, Figure 4.2, each node corresponds to a vertex in the graph. The vertices are joined with an edge if the corresponding nodes are connected by an element. For these types of problems, the vertices represent the solution points and the edges show the interactions between them. The information provided by a nodal graph is sufficient to perform initial partitioning, repartitioning, and equation numbering.

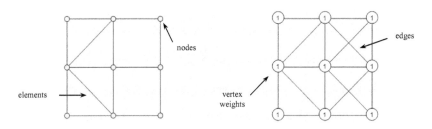

Figure 4.2 The Real Structure and Its Nodal Graph

The serial CSR (compressed storage) [30] format is used to store the data which describes nodal graph. In this format, the vertex adjacency information is stored in two arrays. The first array keeps the beginning address of the adjacent vertex list of the ith vertex. The adjacent vertices of ith vertex are kept in the second array. The vertex weights are kept in a third array whose size is equal to the number of vertices. The METIS [30] library utilizes the serial CSR format.

The CSR format is sufficient for initial partitioning which is performed serially; however all the workload balancing computations are performed in parallel. Each processor needs to store not only the vertex adjacency information of their substructures, but also the adjacency information of the interface vertices between adjacent partitions. Thus, subgraphs are developed for this purpose. Each subgraph can be pictured as a nodal

114

graph of a substructure with links to adjacent substructures at interfaces. The subgraph data is stored in distributed CSR format [28], the same format that PARMETIS [28] utilizes.

Both METIS [30] and PARMETIS [28] libraries use an integer vector to store the partitioning information. The jth element of this vector points to the substructure which contains the jth vertex. This is the only data transferred during the workload balancing iterations. The partitioning information obtained during each iteration is stored in an integer matrix having a size of the maximum number of iterations by the number of nodes in the structure.

4.2.3 Initial Partitioning

METIS [30], a multilevel graph partitioning library, was utilized to create the initial partitions. METIS provides two algorithms for partitioning an unstructured graph into k-parts. The first algorithm, called k-way partitioning, directly partitions the coarsened graph into k-parts. Then a k-way refinement algorithm is utilized to project back to the original graph. The second algorithm, called recursive partitioning, obtains the desired number of partitions by recursively dividing the graph and its subsequent subgraphs into two.

The METIS manual [30] recommended using recursive partitioning algorithm for a small number of partitions (up to 32). As the number of partitions increase, the k-way partitioning algorithm becomes considerably faster than the recursive one. In this study, a maximum of 12 processors were used for solution and the recursive partitioning algorithm was chosen as the initial partitioning algorithm.

4.2.4 Data Distribution

Performing workload balancing iterations in parallel has advantages. It allows performing equation numbering, operation count computations of substructures and repartitioning synchronously. Moreover, there will be no change in the elements of the structure throughout the solution during workload balancing iterations. In other words, the nodal graph of the structure will remain the same during the iterations. The only change that occurs is how the structure is partitioned into substructures, or in other words, how the nodal graph is partitioned into subgraphs, as shown in Figure 4.3. Thus, there is no need for vertex migration during the workload balancing iterations.

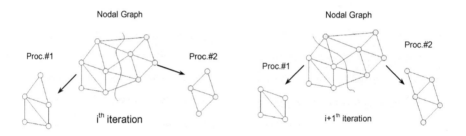

Figure 4.3 Nodal Graph and Subgraphs during Iterations

In order to minimize communication between processors, the nodal graph of the structure is distributed and kept in the local memory of each processor before the iterations begin. When the repartitioning algorithm provides new information about the substructures, only the partitioning information, which is an integer array, is distributed to processors. The subgraphs are extracted from the nodal graph using the partitioning information without the need of any extra communication.

4.2.5 Numbering of Substructure Vertices

The aim of this step is to number the vertices in each substructure in such a way that, when a single degree of freedom is assigned to a vertex, the profile or the bandwidth of the resulting stiffness matrix is minimized. By assigning a single degree of freedom to each vertex, the size of the adjacency information of the substructures are reduced, thus equation numbering is performed much faster. While assigning the degrees of freedom during the stiffness matrix assembly, each vertex is visited one by one according to this numbering and the active degrees of freedom are numbered sequentially.

The vertices are numbered using two different equation reordering algorithms; Gibbs-Poole-Stockmeyer [14] bandwidth minimization and the Gibbs-King [15] profile reduction algorithms. Once the subgraphs for each substructure are prepared, the internal vertices are separated from the interface ones and numbered by using one of the above algorithms. Then, the interface equations are numbered. Since there is no interaction between the numbering of the internal and the interface equations, the average column heights of the interface equations will be much higher than the internal ones and the resulting shape of the stiffness matrix will look like an arrow. Moreover, the column heights of the interface equations are significantly affected by the way the internal equations are numbered. Thus, in order to reduce the column heights of the interface equations, the effect of reversing the numbering of the internal equations is checked and the numbering that produces the lowest operation count is chosen.

4.2.6 Vertex weight definitions

The repartitioning algorithms balance the partitions by equating the sum of the vertex weights of partitions by transferring vertices from one partition to another. Thus, the vertex weights must be defined in such a way that their sums in each partition represent the imbalance in the condensation times.

For that purpose, the imbalance factor of each substructure is first calculated. The imbalance factor shows the relative condensation times of substructures and is calculated with Equation (4.1):

$$I(p) = \left(\frac{OC(p)}{\sum_{i=1}^{n} OC(i)} \right)^{\alpha} \tag{4.1}$$

where p is the substructure id, 'OC' is the operation count value of the pth substructure, 'α' is a scaling factor which will de discussed below, and 'I' is the imbalance factor of the pth substructure. The substructures and processors have the same id's because each processor is responsible for a single substructure. The imbalance factor actually illustrates the workloads of each processor. The smaller its value, the faster is the condensation of the pth substructure.

The scaling factor 'α' is utilized to decrease the difference between the imbalance factors of substructures. For example, when 'α' is taken as 0.5, the square roots of the operation count values are utilized to compute the imbalance factors. Having closer imbalance factors among substructures reduces the node transfers during repartitioning.

In this study, the 'α' value was taken as '1' for structures whose nodes are connected to almost a uniform number of elements. When the nodes are exchanged among the substructures of such structures, the internal equation numbering order usually remain unchanged as shown in Figure 4.4a. If the number of elements connected to a node varies extensively in a structure, adding or removing nodes from the structure's substructures will more likely change the internal equation numbering order as shown in Figure 4.4b. When the internal equation numbering order changes, the column heights of the interface equations change in an unpredictable way that may drastically affect the condensation time since the internal and interface equations are numbered independently. As a result, the iterations usually do not converge to a balanced solution and the governing condensation time of substructures is usually not decreased. Thus, the structures whose nodes are connected to varying number of elements were considered to be very sensitive to modifications and the workload balancing iterations were performed by taking 'α' equal to 0.5.

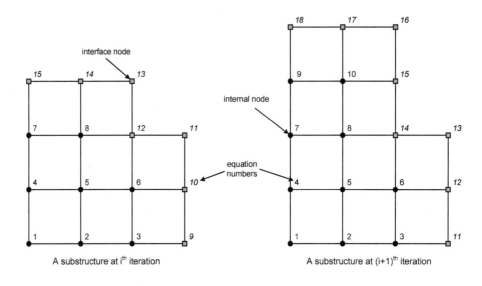

(a) Internal numbering order remains unchanged

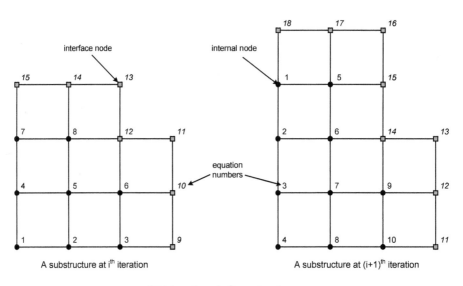

(b) Internal numbering order changes

Figure 4.4 Equation Numbering of Substructures Before and After Modifications

The vertex weights are calculated using the imbalance factors of substructures. The same weight values are assigned to the vertices belonging to the same substructure and are calculated with Equation (4.2):

$$Wu(p) = nT \cdot \frac{I(p)}{n(p)} \tag{4.2}$$

where 'nT' s the number of nodes in the structure, 'n' is the number of vertices in the pth substructure and 'Wu' is the vertex weight that will be assigned to all vertices of the pth substructure. In this equation, the total weight that will be assigned to a substructure is calculated by multiplying the imbalance factor of the pth partition with the total number of nodes in the structure. The weight that will be assigned to each vertex is found by simply dividing that value with the number of vertices in the pth partition.

4.2.7 Repartitioning

The PARMETIS [28] library includes various repartitioning algorithms of both the diffusion and scratch-remap type. The diffusion algorithms are based on the multilevel scheme as discussed in Section 2.2.1. During the coarsening phase, only the vertices belonging to the same partition are merged. This way, the coarsened partitions are identical to the original ones. The workload diffusion, either local [41] or global [25], is performed on these coarsened graphs. Once the graph is balanced, the multilevel refinement begins. The interface vertices are visited randomly and checked if they can improve the balance or decrease the edge-cut by migrating the vertices to the adjacent substructure. If so, the interface vertices are migrated.

The PARMETIS [28] library also provides two different versions of the scratch-remap algorithms. The basic algorithm uses the original graph for repartitioning and remapping whereas the multilevel algorithm performs computations at the coarsened level.

The study by Schloegel et al. [39] compared the performances of the diffusion and scratch-remap algorithms of PARMETIS [28] on generated 3D adaptive meshes. The scratch-remap algorithms produced lower edge-cut and similar vertex migration volumes than the diffusion algorithms for problems with a large amount of localized imbalances. For other types of problems, they favored diffusion.

In the examples tested in this study, the effects of both local diffusion and scratch-remap type repartitioning algorithms on the resulting substructures were investigated.

4.2.8 Convergence Criteria and Choosing Partitioning for Solution

The ultimate goal of this method is to increase the efficiency of the parallel substructure solver by decreasing the time spent during condensation as much as possible. Since the condensation time is governed by the slowest substructure, the presented method attempts to better balance the condensation time by transferring nodes from substructures with higher estimated condensation times to the ones with lower estimated condensation times. By doing this, it is assumed that balancing the condensation times of the substructures will reduce the governing condensation time.

On the other hand, obtaining well-balanced substructures depends highly on the irregularity of the structure. For some structures, it is not possible to say a balanced solution exists for the given number of substructures. Moreover, there may be some cases where even the algorithm converges to a balanced solution, one of the partition created during the iterations produced lower condensation times. Therefore, all the partitioning informations obtained during the iterations are stored. Once the iterations are completed, the partitioning that has the fastest estimated solution time is selected for the actual partitioning. The solution time estimations can consider the condensation time only or the total solution time including the interface solution time estimation.

During the iterations, as the sum of vertex weights of the partitions become closer to each other, the changes in the partitions decrease. Hence, after each repartitioning, the algorithm checks whether any new modifications occurred in the partitions. If not, it is assumed that the solution has converged and the workloads of substructures are balanced. The iterations stop at this point. If the partitions are modified after the repartitioning, the iterations continue until the maximum number of iterations has been reached. The results of the example problems showed that the best results were usually obtained before 20 iterations. Thus, the maximum number of iterations was taken as 20.

4.2.9 Interface Element Assignment

Up to this step, the structure is represented as a nodal graph and this graph is utilized in all computations. When the nodal graph is partitioned or repartitioned, the results will indicate to which partition a vertex belongs, but there will be no information about the elements at the interfaces as shown in Figure 4.5. During the iterations, the algorithm ignores the effect of interface elements in order to simplify the problem by not adding another heuristic approach to the iterations. Hence, it is assumed that the condensation time improvement obtained without the effects of interface elements will be close to the improvement obtained with them.

123

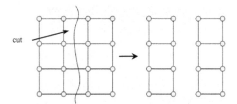

Figure 4.5 The Actual Structure and Substructures Used during Iterations

The interface elements have nodes on two or more substructures. While preparing the structural data from the partitioning information, a decision must be made in order to assign each interface element to a substructure. When a new interface element is assigned to a substructure, new nodes are created, as shown in Figure 4.6. This addition will affect the condensation time for that substructure.

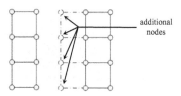

Figure 4.6 Substructures with Interface Elements

The interface element assignment algorithm attempts to distribute the interface elements to the substructures based on two criteria: keeping the number of interface nodes for each substructure as low as possible and assigning more elements to the substructures having lower operation counts for condensation. In order to assign the interface elements using these criteria, each interface element is investigated in a random order and a cost value is calculated by using Equation (4.3):

$$ct(p) = I(p) \cdot (ne - np) \qquad (4.3)$$

where 'ne' is the number of nodes in the element and 'np' is the number of the element's nodes in the pth substructure.

The equation consists of two parts. The second part, '(ne-np)' forces the algorithm to assign the interface element to the substructure which has more nodes. If an element has an equal number of nodes in two or more substructures, the first part of the equation governs. For such conditions, the element is assigned to one of the adjacent substructures that has the lowest computational load. In other words, the element is assigned to the substructure which has the smallest cost value calculated according to Equation (4.3).

4.3 Illustrative Examples and Discussions

4.3.1 2D Square Mesh Model

4.3.1.1 Substructures during Iterations

The following example illustrates the change in the substructures during iterations with the diffusion and scratch-remap algorithms. The shape of the partitioning, the improvement ratios and the SS Ratios are shown for each iteration where the SS Ratio represents the difference between the maximum and minimum condensation times over all the substructures at each iteration. Its value is computed by Equation (4.4):

$$SSRatio = \frac{max(OC(p)) - min(OC(p))}{min(OC(p))} \qquad (4.4)$$

where 'OC(p)' is the operation count for condensation of pth substructure. The smaller the SS Ratio, the more balanced are the substructures.

The improvement ratio shows the improvement of the governing condensation time of the current partitioning when compared with the initial governing condensation time. The smaller its value, the faster are the condensations. It is calculated by Equation (4.5).

$$Ip(i) = \frac{max(OC(i, p))}{max(OC(1, p))} \qquad (4.5)$$

In the above equation, the improvement ratio, 'Ip', is found by dividing the maximum condensation time of substructures at the ith iteration by the maximum condensation time of the initial substructuring.

Workload Balancing with Diffusion Algorithm

Figure 4.7 illustrates the substructures created using the diffusion algorithm. For this example problem, a 2D square mesh having 25,600 quadrilateral elements was partitioned into 8 substructures. During the computation of nodal weights, the imbalance scaling factor, 'α' was taken as 1.

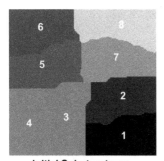

Initial Substructures
SS Ratio: 2.44
Improvement: 1.00

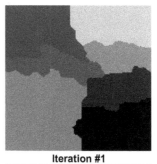

Iteration #1
SS Ratio: 1.43
Improvement: 1.06

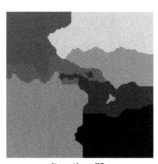

Iteration #2
SS Ratio: 1.03
Improvement: 1.02

Iteration #3
SS Ratio: 1.28
Improvement: 0.91

Iteration #4
SS Ratio: 0.80
Improvement: 0.96

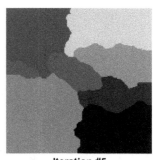

Iteration #5
SS Ratio: 0.32
Improvement: 0.73

Figure 4.7 Substructures at Each Iteration, Diffusion Algorithm

127

Iteration #6
SS Ratio: 0.28
Improvement: 0.67

Iteration #7
SS Ratio: 0.39
Improvement: 0.75

Iteration #8
SS Ratio: 0.19
Improvement: 0.71

Iteration #9
SS Ratio: 0.35
Improvement: 0.78

Iteration #10
SS Ratio: 0.20
Improvement: 0.67

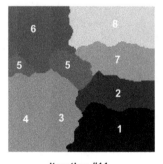

Iteration #11
SS Ratio: 0.21
Improvement: 0.67

Figure 4.7 Substructures at Each Iteration, Diffusion Algorithm (cont.)

The initial partitioning contained an equal number of nodes in each substructure; but the SS Ratio value was 2.44. In other words, there was 2.44 times difference in the condensation times for the fastest and the slowest substructure. This occurred because the second, third, fifth and seventh substructures required more condensation time since they contained more interface nodes than the others. The first two iterations did not provide any speed-up in the condensation times although the SS Ratio dropped. After iteration two, the governing condensation time of the substructures started to become faster than the governing condensation time of the initial substructures.

The algorithm converged at the eleventh iteration. The repartitioning algorithm did not provide any changes in the partitioning at the twelfth iteration, thus the iterations were finalized.

When the final substructures are compared with the initial ones, the number of internal nodes was decreased for the second, third, fifth and seventh substructures but their number of interface nodes remained almost the same. In other words, their shapes did not change much but their edges approached each other. The fifth substructure was split into two parts at the fourth iteration and remained the same way as the iterations were finalized. The final substructures had much more balanced condensation times as evidenced by an SS Ratio of 0.21, and the governing condensation time decreased by 33%.

Workload Balancing with Scratch-Remap Algorithm

In this example, the same structure was balanced using the scratch-remap repartitioning algorithm. The shapes of substructures at each iteration are presented in Figure 4.8. Similar to the previous run, the 'α' factor was taken as 1.

The algorithm converged after 10 iterations. The final SS Ratio value was 0.33 which was slightly worse than SS Ratio of the final substructures created with the diffusion algorithm. On the other hand, the decrease in the governing condensation time improved around 37%. Thus, balancing the condensation times of the substructures decreases the governing condensation time, but having more balanced condensation times does not always mean that the governing condensation time is faster. There may be another partitioning which the substructures are less balanced but the governing condensation time is faster.

The final substructures created with the scratch-remap algorithm had more uniform shapes than the ones created with the diffusion algorithm. First of all, none of the substructures were split into more than one part. Secondly, there were four similar substructures at the corners having larger number of internal nodes. The second and the fifth substructures were located at the center of the mesh. They had the least number of nodes because all their edges are at the interfaces. The other two midsize substructures, the third and the seventh, have interface nodes at their three sides. As a result, the iterations resulted in a decrease in the number of nodes of the substructures which contained more interface nodes.

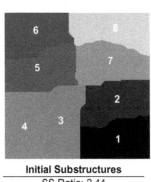

Initial Substructures

SS Ratio: 2.44
Improvement: 1.00

Iteration #1

SS Ratio: 1.06
Improvement: 0.70

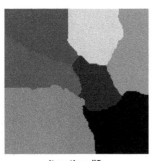

Iteration #2

SS Ratio: 0.61
Improvement: 0.74

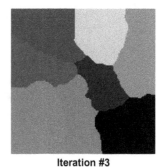

Iteration #3

SS Ratio: 0.48
Improvement: 0.70

Iteration #4

SS Ratio:1.28
Improvement: 0.82

Iteration #5

SS Ratio: 0.75
Improvement: 0.91

Figure 4.8 Substructures at Each Iteration, Scratch-remap Algorithm

Iteration #6

SS Ratio: 0.82

Improvement: 0.78

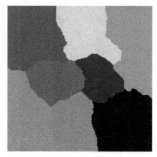

Iteration #7

SS Ratio: 0.66

Improvement: 0.82

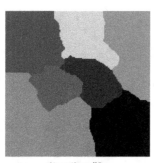

Iteration #8

SS Ratio: 0.61

Improvement: 0.84

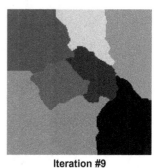

Iteration #9

SS Ratio: 1.23

Improvement: 0.76

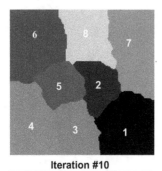

Iteration #10

SS Ratio: 0.33

Improvement: 0.63

Figure 4.8 Substructures at Each Iteration, Scratch-remap Algorithm (cont.)

4.3.1.2 Effect of Interface Element Assignment

The next two tables, Tables 4.1 and 4.2, show the operation counts for condensation and the number of nodes of substructures after the iterations and with the interface elements. The last two columns of the tables show the difference between them. The change in the operation count for a substructure is calculated by dividing the operation count value for the substructures with interface elements with the operation count value for substructures after iterations. The first table, Table 4.1, is for the final substructures of the 2D square mesh balanced with the diffusion algorithm whereas the second table, Table 4.2, is for the substructures balanced with the scratch-remap algorithm.

The SS Ratios of the final substructures were decreased for both cases which indicated that the balancing strategy during the interface element assignment was successful. The increase in the operation counts was around 10% except the sixth substructure in Table 4.1. It had two distinct structures that caused it to share more interface nodes with its adjacent substructures. Moreover, it shares an interface with six other substructures. Since it had the smallest operation count value, most of the interface elements were assigned to it. As a result, it had 185 additional nodes with a 25% increase in the operation count.

As can be seen from Table 4.1, the operation count for the governing condensation time was 5.13×10^7 after the iterations. When the interface elements were assigned to the substructures, the operation count for the governing condensation time increased to 5.38×10^7. Similarly, the operation count for the governing condensation time increased from 4.82×10^7 to 5.02×10^7 for substructures balanced with the scratch-remap algorithm.

For both cases, the governing condensation times were increased by only 5% due to the assignment of the interface elements.

Table 4.1 Operation Count Values Before and After Interface Elements Assignment for
Substructures Balanced with Diffusion Algorithm

| Processor Id | Substructures After Iterations | | Substructures with Interface Elements | | Difference | |
	Operation Count	Number of Nodes	Operation Count	Number of Nodes	Change in Operation Count	Additional Nodes
1	4.87E+07	4308	5.14E+07	4410	105%	102
2	5.03E+07	2693	5.38E+07	2761	107%	68
3	4.77E+07	2275	5.32E+07	2444	112%	169
4	4.84E+07	3984	5.31E+07	4063	110%	79
5	5.13E+07	4095	5.31E+07	4157	104%	62
6	4.27E+07	1756	5.31E+07	1941	124%	185
7	4.82E+07	2790	5.25E+07	2909	109%	119
8	4.46E+07	4020	4.77E+07	4157	107%	137
	SS Ratio: 0.201		SS Ratio: 0.129			

Table 4.2 Operation Count Values Before and After Interface Elements Assignment for
Substructures Balanced with Scratch-Remap Algorithm

| Processor Id | Substructures After Iterations | | Substructures with Interface Elements | | Difference | |
	Operation Count	Number of Nodes	Operation Count	Number of Nodes	Change in Operation Count	Additional Nodes
1	4.25E+07	3962	4.65E+07	4050	110%	88
2	4.49E+07	2046	5.02E+07	2143	112%	97
3	3.62E+07	2508	4.03E+07	2654	111%	146
4	3.74E+07	3727	3.98E+07	3805	107%	78
5	4.74E+07	4296	4.81E+07	4331	102%	35
6	4.18E+07	2017	4.72E+07	2141	113%	124
7	4.82E+07	4381	4.90E+07	4420	102%	39
8	4.44E+07	2984	4.99E+07	3157	112%	173
	SS Ratio: 0.330		SS Ratio: 0.261			

4.3.2 Three Story Building

4.3.2.1 Substructures Before and After Iterations

The workload balancing algorithm successfully balanced the condensation times of the 2D square mesh. It decreased the SS Ratio from 2.44 to 0.21 and 0.33 using the diffusion and scratch-remap repartitioning algorithms, respectively. The governing condensation time decreased approximately 33% for both cases. However, the 2D square mesh problem had a very uniform shape. It was symmetric and composed entirely of 2D quadrilateral elements. The resulting stiffness matrix had an almost constant column height, and mostly the same number of elements was joined to the nodes except for the nodes on the outside boundary. However, many civil engineering structural models have more irregular shapes and are composed of a combination of frame and 2D shell elements.

In order to see determine if the balancing algorithm provided better governing condensation time than the governing condensation time of the initial substructures for a less uniform structure, a three story building model shown in Figure 4.9 was tested.

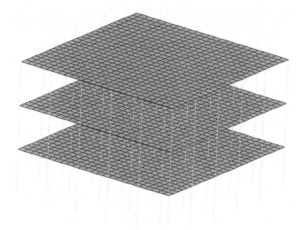

Figure 4.9 The Initial Structure

The structure was first partitioned into 8 substructures using the METIS recursive bisection partitioning algorithm. The initial substructures are presented in Figure 4.10. The SS Ratio was computed to be 0.67 and the condensation was governed by the fifth substructure.

The workload balancing iterations were first performed by using the scratch-remap algorithm with 'α' equal to 1. The iterations did not provide successful partitioning for this case. The governing condensation times obtained during the iterations were slower than the governing condensation time of the initial substructures. In the 2D square mesh example, as the nodes are transferred among the substructures, the average internal column height and the number of equations of their stiffness matrix were changed, but the numbering order of the internal nodes remained almost the same. Due to this, the changes in the condensation times after each repartitioning were small.

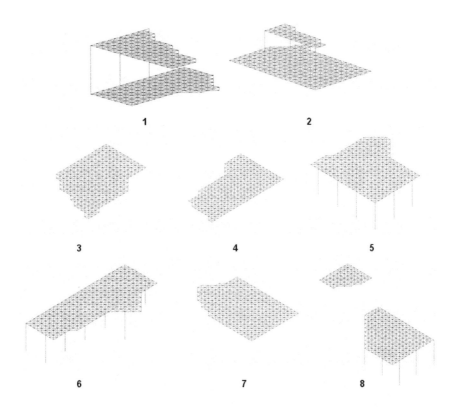

Figure 4.10 Initial Partitions Created by METIS Recursive Bisection Partitioning
Algorithm

137

On the other hand, in this example the modifications at the substructures also altered their internal node numbering order. Since the internal and interface equations are numbered independently, such alterations significantly affected the average column height at the interfaces. Thus, the changes in the condensation times due to node transfers were much more significant than in the previous example. In other words, the substructures of the 3 story building model were more sensitive to modifications.

In order to prevent large changes in the substructures during the iterations, the same problem was solved again using the 'α' value set to 0.5. This time, the iterations converged after 3 iterations. Since the square roots of the operation counts are balanced, the actual condensation times of the final substructures would be less balanced. The final substructures of the three story building are shown in Figure 4.11.

The final substructures were similar to the initial ones. The SS Ratio dropped to 0.21 and the governing condensation time decreased by approximately 20%. This time, the seventh substructure had the slowest condensation time. The fifth substructure transferred some of its vertices to the sixth substructure and similarly the second substructure transferred its vertices to the fourth substructure. As a result, using a smaller 'α' value decreased the changes in the substructures and the balanced solution was obtained with minor modifications to the initial geometry.

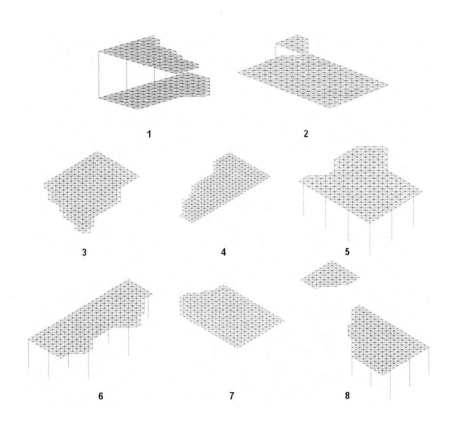

Figure 4.11 Partitions Balanced with Scratch-Remap Repartitioning Algorithm

4.3.2.2 Effect of Interface Element Assignment

Figures 4.12 to 4.19 present the balanced substructures for the 3 story building model before and after interface element assignment. The figures on the left are the 3D representation of the nodal subgraph used during the iterations whereas the figure on the right is the structural model used during the condensation. The table at the bottom of each figure shows the operation counts and the number of nodes before and after the interface element assignment and their differences. The change in operation count for a substructure is calculated by dividing the operation count value for the substructures with interface elements with the operation count value for the substructures after the iterations have converged.

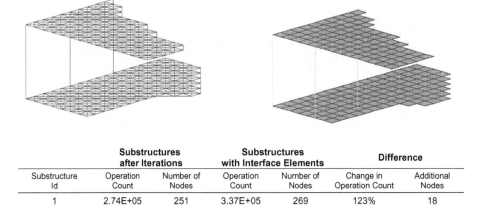

	Substructures after Iterations		Substructures with Interface Elements		Difference	
Substructure Id	Operation Count	Number of Nodes	Operation Count	Number of Nodes	Change in Operation Count	Additional Nodes
1	2.74E+05	251	3.37E+05	269	123%	18

Figure 4.12 First Substructure Before and After Interface Element Assignment

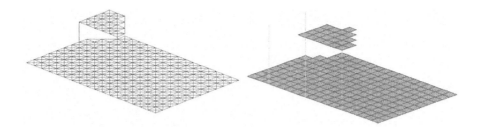

	Substructures after Iterations		Substructures with Interface Elements		Difference	
Substructure Id	Operation Count	Number of Nodes	Operation Count	Number of Nodes	Change in Operation Count	Additional Nodes
2	2.62E+05	235	3.02E+05	259	115%	24

Figure 4.13 Second Substructure Before and After Interface Element Assignment

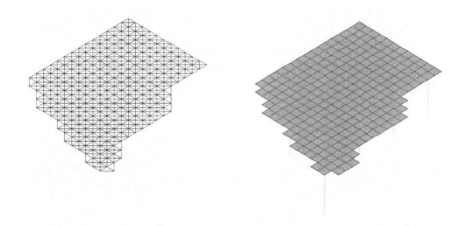

	Substructures after Iterations		Substructures with Interface Elements		Difference	
Substructure Id	Operation Count	Number of Nodes	Operation Count	Number of Nodes	Change in Operation Count	Additional Nodes
3	2.32E+05	264	3.22E+05	313	139%	49

Figure 4.14 Third Substructure Before and After Interface Element Assignment

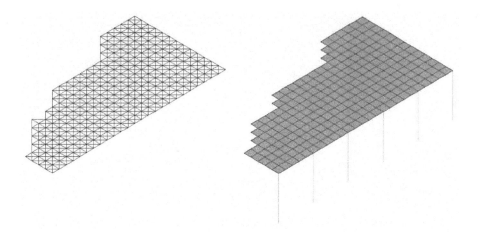

Substructure Id	Substructures after Iterations		Substructures with Interface Elements		Difference	
	Operation Count	Number of Nodes	Operation Count	Number of Nodes	Change in Operation Count	Additional Nodes
4	2.53E+05	273	2.92E+05	299	115%	26

Figure 4.15 Fourth Substructure Before and After Interface Element Assignment

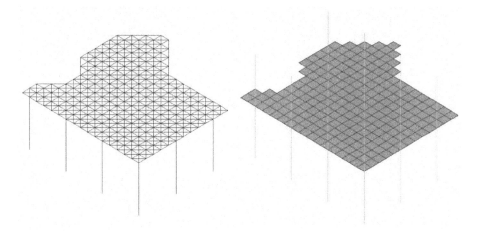

Substructure Id	Substructures After Iterations		Substructures with Interface Elements		Difference	
	Operation Count	Number of Nodes	Operation Count	Number of Nodes	Change in Operation Count	Additional Nodes
5	2.28E+05	244	2.47E+05	286	108%	42

Figure 4.16 Fifth Substructure Before and After Interface Element Assignment

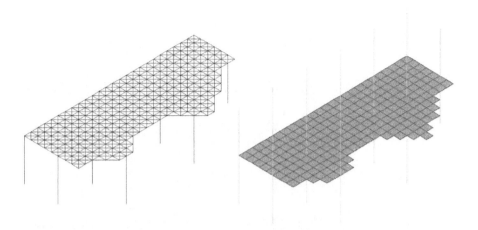

	Substructures after Iterations		Substructures with Interface Elements		Difference	
Substructure Id	Operation Count	Number of Nodes	Operation Count	Number of Nodes	Change in Operation Count	Additional Nodes
6	2.71E+05	262	3.23E+05	292	119%	30

Figure 4.17 Sixth Substructure Before and After Interface Element Assignment

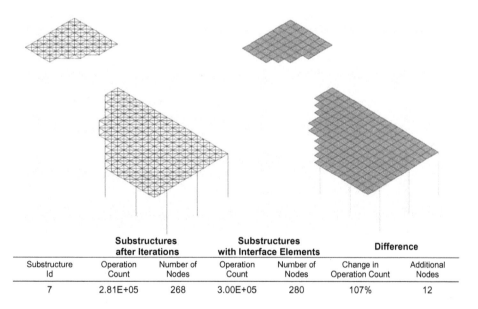

	Substructures after Iterations		Substructures with Interface Elements		Difference	
Substructure Id	Operation Count	Number of Nodes	Operation Count	Number of Nodes	Change in Operation Count	Additional Nodes
7	2.81E+05	268	3.00E+05	280	107%	12

Figure 4.18 Seventh Substructure Before and After Interface Element Assignment

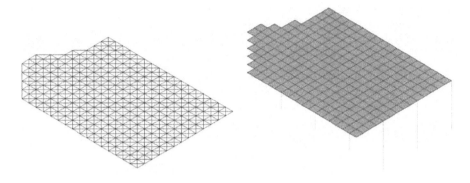

	Substructures After Iterations		Substructures with Interface Elements		Difference	
Substructure Id	Operation Count	Number of Nodes	Operation Count	Number of Nodes	Change in Operation Count	Additional Nodes
8	2.24E+05	267	2.80E+05	310	125%	43

Figure 4.19 Eighth Substructure Before and After Interface Element Assignment

The third, fifth, and the eighth substructures received more additional nodes during the interface element assignment due to their operation counts after iterations being the smallest. The least number of interface elements were assigned to the seventh substructure which had the maximum operation count before the assignment. The highest increase was observed for the third substructure, by 39%, followed by the first one. Although only eighteen additional nodes were added to the first substructure, the operation count increased by 23%. On the other hand, although the fifth substructure had 42 additional nodes, the operation count increased by only 8%. This was because the additional nodes changed the internal numbering order of the fifth substructure which decreased the column heights of the interface equations. Therefore, the increase in the condensation time for the fifth substructure was much smaller when compared with the other ones.

4.4 Example Problems and Discussions

Various example problems were solved on DELL cluster using from 2 to 12 computers. For each example problem, the substructures were created by either the diffusion or the scratch-remap type repartitioning algorithm and compared with the results of the initial substructuring. Moreover, for some problems, the effect of using bandwidth minimization or profile reduction algorithms was investigated. The first three example problems, '2D Square Mesh', 'Half Disk', and 'Bridge Deck' are uniform structures. They are composed of single element type and have almost uniform element connectivity. Moreover, they are mid-size problems; in other words, their serial solution time is less than an hour and they can be solved with an in-core condensation algorithm. The other three examples, 'High-rise Building I & II' and "Nuclear Waste Plant', are composed of a combination of frame and shell elements. The condensation times for the substructures of these examples are more sensitive to modifications made during the repartitioning.

Two graphs are presented for each problem. The graph on the left shows the condensation times for the 2 to 12 processor solutions. The graph on the right shows the speed-ups, i.e. how much the condensation time decreased as the number of processors increased. It is calculated by dividing the estimated serial solution time by the governing condensation time for a given number of processors, Equation (4.5). The differences between the condensation times of different partitionings are better differentiated on the speed-up graph.

$$Speed - up = \frac{Serial\ Solution\ Time}{Parallel\ Condensation\ Time} \qquad (4.5)$$

4.4.1 2D Square Mesh

The 2D Square Mesh model was solved by using 2 to 12 computers in order to see the improvement obtained by the workload balancing algorithm. The substructures were created by using both diffusion and scratch-remap algorithms and their results were compared with the results of the initial substructures. Shell elements were assigned to each element in the mesh; hence each node had 6 degrees of freedom. For all cases, the equations were numbered with the bandwidth minimization algorithm and the 'α' value was taken as 1.

Table 4.3 presents the actual condensation times for the 2D square mesh with 8 processors using the initial substructures and the substructures balanced using the diffusion and scratch-remap algorithms. The actual operation count values in Table 4.3 are 6^3 times more than the operation counts used during the iterations since each node has 6 degrees of freedom.

The condensation time using the initial substructures was governed by the seventh substructure and was equal to 101.91 seconds. The maximum difference between the condensation times was around 70 seconds which validates the significant imbalance calculated using the SS Ratio as shown in Figure 4.7. As can be observed in Table 4.3, the condensation times of the substructures balanced using the diffusion algorithm were

146

much more balanced. The maximum difference between the condensation times dropped to 8 seconds. The governing condensation time also decreased to 70.42 seconds which was 30% less than the one with the initial substructures. The substructures balanced with the scratch-remap algorithm were faster in terms of governing condensation time but the substructures were less balanced.

Table 4.3 Condensation Times of 2D Square Mesh on DELL Cluster

	Operation Counts			Condensation Times		
Proc. Id	Initial	Diffusion	Remap	Initial	Diffusion	Remap
1	5.986E+09	1.125E+10	1.020E+10	35.97	67.83	61.69
2	1.415E+10	**1.175E+10**	**1.095E+10**	84.74	**70.42**	**65.47**
3	1.379E+10	1.161E+10	8.810E+09	82.52	69.60	52.75
4	5.282E+09	1.161E+10	8.735E+09	31.65	69.57	52.44
5	5.519E+09	1.164E+10	1.055E+10	33.25	70.33	63.79
6	1.504E+10	1.157E+10	1.031E+10	89.98	69.31	61.75
7	**1.709E+10**	1.146E+10	1.073E+10	**101.91**	68.60	64.32
8	6.620E+09	1.045E+10	1.091E+10	39.78	62.66	65.29
Max:	1.709E+10	1.175E+10	1.095E+10	101.91	70.42	65.47

The other important aspect of the workload balancing step is the time required for the iterations. Table 4.4 shows the time spent during iterations for both the diffusion and scratch-remap repartitioning algorithms. The substructures were balanced after 11 iterations with the diffusion algorithm and 10 iterations with the scratch-remap algorithm. The time spent during each iteration was around 0.30 seconds. As can be seen from Table 4.4, approximately 4 seconds was spent for workload balancing using the scratch-remap algorithm which resulted in a 36 second (36%) reduction in the actual condensation time.

147

Table 4.4 Time Spent During Iterations

Repartitioning Algorithm	Initial Partitioning	Time spent during Iterations	Total
Diffusion	0.56 secs	3.45 secs	4.01 secs
Scratch-Remap	0.53 secs	3.16 secs	3.67 secs

Figure 4.20 presents the actual condensation times and the speed-up obtained when 2 to 12 processors were utilized for solution. The initial substructures for solutions with 2 and 4 processors were balanced. After that point, a significant imbalance occurred due to an unequal number of interface nodes between the substructures. For example, the SS Ratio was 1.87 for 6 substructures initially and the governing condensation time was slower than the one with 4 processors. Both balancing methods were able to decrease the SS Ratio and decreased the condensation time. The SS ratio reduced to 0.47 and 0.40 and the condensation time decreased by 30% and 33% with diffusion and scratch-remap algorithms, respectively for solution with 6 processors. For other cases, the decrease in condensation time remained between 30% to 42% with both algorithms. Overall, the scratch-remap based balancing produced faster governing condensation times than the diffusion based balancing.

As shown at the speed-up graph of Figure 4.20, when the mesh was divided into four substructures, the condensation time was decreased by 6 times. Likewise, 8 substructures produced a 13 times decrease and 12 substructures produced a 24 times decrease in the governing condensation time. This is mainly because the substructuring not only decreased the size but also the column height of the stiffness matrix of substructures. Thus, when a structure is partitioned into 'p' substructures, the maximum reduction that could be obtained in the condensation time is approximately equal to 'p^3' since the condensation time is a function of 'nm^2' (Equation 2.15) where 'n' is the number of internal equations and 'm' is the average column height of internal equations.

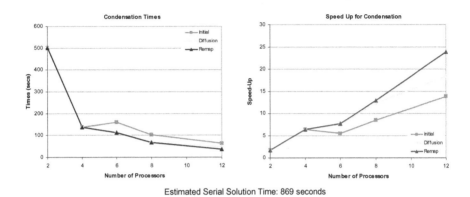

Estimated Serial Solution Time: 869 seconds

Figure 4.20 Condensation Times for 2D Square Mesh

4.4.2 Half Disk

The Half Disk model, Appendix A.1.2, was first solved on DELL cluster to determine the effects of diffusion and scratch-remap based balancing on the condensation times. The equations were numbered with profile reduction algorithm and the 'α' value was taken as 1. Figure 4.21 shows the difference between the initial and the balanced substructures.

The balanced substructures produced much faster condensation times than the initial ones for all parallel solutions. The decrease in the condensation time ranged between 30% to 50%. Similar to the previous example problem, the scratch-remap algorithm performed better than the diffusion algorithm; it produced the fastest condensation times. The time spent during the iterations was around 9 seconds which was not significant

149

when compared with the condensation time. The iteration time remained almost constant as the number of processors increased because the cost of increased communication during repartitioning was balanced by the decrease in the time spent during equation numbering.

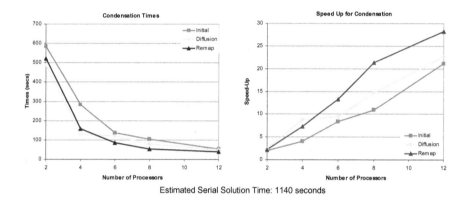

Estimated Serial Solution Time: 1140 seconds

Figure 4.21 Condensation Times for Half Disk

The second series of runs were performed to investigate the effects of different numbering algorithms on the resulting substructures. Figure 4.22 shows the differences between the bandwidth minimization and profile reduction algorithms. For both cases, the scratch-remap algorithm was used for repartitioning.

As can be seen from Figure 4.22, the difference between the condensation times was insignificant between the substructures created with both numbering algorithms The substructures created with the profile reduction algorithm outperformed the substructures created with the bandwidth reduction algorithm with 6 processors whereas the situation is the opposite with 12 processors. As a result, it is not possible to claim the superiority of one numbering algorithm over the other.

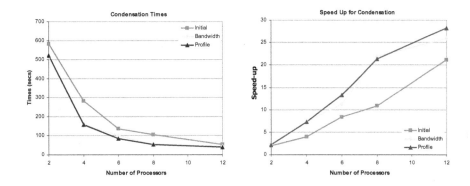

Figure 4.22 Condensation Times for Half Disk

4.4.3 Bridge Deck

The Bridge Deck model, Appendix A.1.3, is another uniform structure. It is composed entirely of brick elements and has a symmetric geometry. The substructures were balanced with the diffusion algorithm by using both bandwidth minimization and profile reduction algorithms. Similar to the previous examples, the 'α' value was taken as 1.

Figure 4.23 shows the condensation times of the initial and the balanced substructures. The initial partitioning produced balanced substructures for 4 processor solution, but the condensation times of substructures were highly imbalanced for other cases. The workload balancing algorithm successfully decreased the condensation times for all parallel solutions. The substructures created with the bandwidth minimization algorithm produced better condensation times than the substructures created with the profile reduction algorithm for 6 and 12 processor solutions but the situation was the opposite for 2 and 8 processor solutions. Therefore, similar to the previous example, it is not possible to claim the superiority of one numbering method over the other.

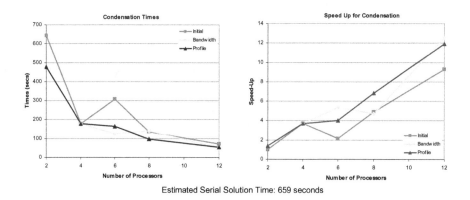

Estimated Serial Solution Time: 659 seconds

Figure 4.23 Condensation Times for Bridge Deck

The speed-up values obtained for this model were not as high as in the previous examples. This was mainly due to the shape of the model. Due to its rectangular-prism like shape as shown in Figure 4.24a, the average column height of the resulting stiffness matrix depended on the connectivity of the nodes through the deck section, Figure 4.24b. Thus, when the deck was divided into two parts, the shape of the section did not change, as can be seen in Figure 4.24c. As a result, this division caused a decrease in the number of nodes but did not affect the average column height and the speed-up values were equal to or less than the number of processors.

(a) The model

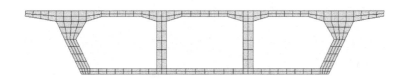

(b) Deck section

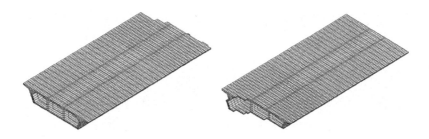

(c) Substructures when the model is partitioned into two

Figure 4.24 Bridge Deck Model

153

4.4.4 High-rise Building I & II

The following example problems are the actual building models that are being constructed in the United States and are shown in Appendix A.1.4 and Appendix A.1.5. They are very large models, each having approximately 328,000 equations. The structural models are composed of a combination of frame, triangular and quadrilateral shell elements. The in-core memory was exceeded for solutions using up to six computers, thus out-of-core solvers were utilized. However, the results in Figures 4.26 and 4.27 are based on in-core condensation time estimations for comparison purposes.

The substructures were created with both diffusion and scratch-remap based balancing algorithms for the first building and the condensation times were compared with the initial substructures. The equations were numbered with the profile reduction algorithm.

As shown in Figure 4.25, the structure has many shear walls and columns. As a result, the structure has highly irregular adjacency structure, in other words the number of elements joined to any node may vary significantly. When the workload balancing iterations were performed by using 'α' value set to 1, the iterations did not converge to a balanced solution and the reductions obtained in the governing condensation time were not high, approximately 15% with 6 substructures, 5% with 8 and 12 substructures. In order to obtain a higher reduction in the condensation time, the 'α' value in Equation (4.1) was taken as 0.5 and much better results were achieved (reduction of 15% with 8 and 12 processors).

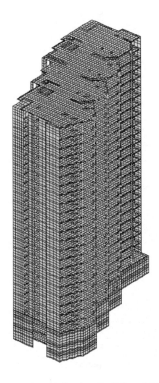

Figure 4.25 High Rise Building I Model

Figure 4.26 presents the condensation times for the 'High-rise Building I' model. The balanced substructures were able to decrease the condensation times in the range of 15% to 30% when compared with the condensation times of the initial substructures. For example, during the solution with 4 processors, the balanced substructures finished the condensations 900 seconds faster than the partitioning based on the initial substructures. The workload balancing step required only 4 seconds. Thus, the time spent during workload balancing iterations when compared with the reduction in the condensation time was insignificant.

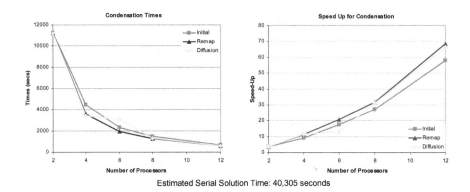

Estimated Serial Solution Time: 40,305 seconds

Figure 4.26 Condensation Times for High-rise Building I

The results of substructures balanced with the scratch-remap algorithm were slightly better than the substructures balanced with the diffusion algorithm. Moreover, for the solution with 6 processors, the governing condensation time using the initial substructures was faster than the governing condensation time of substructures balanced with the diffusion algorithm. This was mainly due to a considerable increase in the condensation time of one of the substructures during interface element assignment.

Figure 4.27 shows the results for the second high-rise building model which has more geometrical irregularities. This time, the initial substructures were compared with the substructures balanced with scratch-remap algorithm. The 'α' value was set to 0.5.

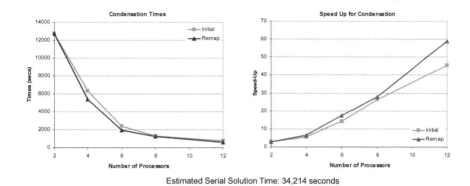

Estimated Serial Solution Time: 34,214 seconds

Figure 4.27 Condensation Times for High-rise Building II

As can be seen from Figure 4.27, the governing condensation time was decreased for all cases except for the solutions with 2 and 8 processors. The improvement in condensation times was 1100, 200 and 100 seconds for solutions with 4, 6, and 12 processors, respectively. The initial substructures were balanced for 2 processors. On the other hand, the workload balancing algorithm was able to decrease the condensation time by only 4% for the 8 processor solution.

4.4.5 Nuclear Waste Plant

The last example problem is the Nuclear Waste Plant model, Appendix A.1.6, which has a very uniform and symmetric geometry; however, the box at the bottom of the model is actually full of many little rooms. The balanced substructures were created with scratch-remap and profile reduction algorithms by using 'α' equal to 0.5. Figure 4.28 shows the condensation times of initial and balanced substructures for 2 to 12 processor solution on the DELL cluster. The results up to six processors are based on time estimations for in-core condensation.

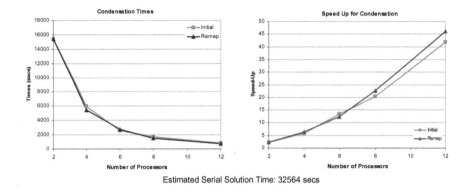

Estimated Serial Solution Time: 32564 secs

Figure 4.28 Condensation Times for Nuclear Waste Plant

The improvements obtained from the iterations ranged between 15% to 25% except for the six processor solution. The improvement in condensation time was around 850, 160 and 70 seconds for 4, 8, and 12 processor solutions, respectively. Overall, the workload balancing step decreased the condensation times more than it consumed.

4.5 Conclusions

The first step of a substructuring based parallel solution method is the division of the structure into smaller substructures. There are various partitioning methods developed for that purpose but none of them is able to balance the condensation times for direct condensation. Variables, such as the number of interface nodes, the column heights at the interfaces etc., may have significant effect on the condensation time but these variables are not considered by the partitioning algorithms. As a result, the initial substructures have highly imbalanced condensation times that reduce the overall performance of the parallel solution.

158

This chapter presented a workload balancing algorithm in order to diffuse the condensation time imbalances by iteratively repartitioning the substructures according to their operation counts for condensation. Two different repartitioning algorithms were utilized during the iterations: diffusion and scratch-remap. The resulting substructures differed according to the type of the repartitioning algorithm. Both algorithms were able to decrease the imbalances in the condensation time of the substructures but the scratch-remap algorithm improved the governing condensation time much better for most cases. Moreover, the substructures created with diffusion algorithm had more interface nodes. Thus, scratch-remap based repartitioning algorithm is a better choice for workload balancing.

During the iterations, each substructure's stiffness matrix was numbered by either the bandwidth minimization [14] or the profile reduction algorithms [15]. The resulting shapes of substructures were affected by the way the equations were numbered. However, according to the results of the test runs, none of the algorithms were superior to each other. Thus, both of them can be utilized.

The scaling factor 'α' has a significant role on the amount of modification occurred at the substructures during repartitioning. For structures whose nodes are connected to almost a constant number of elements, taking 'α' value equal to 1 often produced well balanced substructures and resulted in a considerable decrease in the condensation time. When the number of elements joined to a node in a structure varies extensively, the condensation times of the structure's substructures are highly sensitive to modifications. For such structures, using 1 for 'α' caused vast changes in the estimated condensation time of the substructures after each repartitioning. As a result, more balanced and faster condensation times were not realized. For such cases, the square root of the operation count values were utilized while calculating the imbalance factors by setting 'α' to 0.5. This way, the modifications in substructures occurred during repartitioning were decreased. As a result, an improvement around 20% to 30% in the governing condensation times was obtained.

The other important aspect of this method is the time spent during the iterations. Especially for linear static analysis, the workload balancing step must be finalized very quickly. The results of the example problems showed that the cost of the workload balancing step was insignificant when compared with the improvement in condensation time. The time cost of the iterations does not depend on the size of the problem much but more on the density of the connectivity, the size of the interface and the amount of imbalance. Thus, the presented method is very effective and fast for the linear static analysis.

CHAPTER 5

INTERFACE SOLUTION ALGORITHM

5.1 Introduction

In a substructure based solution method, the substructures are first condensed to the interfaces with other substructures. The condensed substructure stiffnesses are then assembled and the equilibrium equations are solved for the displacements at the interface nodes. Depending on the number of substructures, the interface stiffness matrix may be a very dense matrix. As the number of substructures increases, the density of the interface stiffness matrix decreases. Thus, a parallel interface solver that is capable of efficiently solving both sparse and dense matrices will improve the performance of the solution framework significantly.

The interface problem has been solved with different methods [60, 62, 81, 93]. Farhat et al. [81] implemented the row-wise formulation of the LDL^T method with cyclic row-wise and cyclic block distribution. In their approach, the stiffness matrix was first factorized. Then, by using the lower triangular coefficients of the stiffness matrix, each load vector was factorized. This method could be used for both solving multiple and repetitive load case problem. On the other hand the method required additional data

161

transfers during the forward substitution. Another substructure based solution framework was implemented by Hsieh et al. [89]. They utilized parallel active column solver [93] for the interface problem. They obtained small speed-up during the factorization; and hence have utilized a serial interface solver in their recent work [102]. There are other iterative solvers [62] utilized to solve the interface problem; however, such solvers are not suitable for problems with multiple loading conditions.

In this study, the interface equations are solved by using a parallel version of the LU decomposition method developed for this study. While the active column storage and column-wise factorization were utilized for the condensation step, the variable band storage and row-wise factorization were chosen for the interface solution for the following reasons:

- Normally, the variable band storage requires extra storage in order to keep the upper triangular part of the stiffness matrix when compared with the active column version. However, the row-wise variable band factorization scheme has been shown to outperform active column solvers for dense matrices [78].

- The parallel row-wise factorization has significant advantages during forward and back substitution when compared with the column-wise version. In the column wise approach, the elements of the ith column needs to be transferred to other processors during load factorization [79, 93]. This significantly slows down the solution for systems having slow communication speeds. However, in row-wise factorization, the factorization and forward substitution can be combined and the same data can be utilized for both computations. Moreover, during back

162

substitution, only the ith loading value needs to be transferred. This reduces the communication volume significantly.

5.2 Parallel Variable Band Solver

5.2.1 Serial Implementation

The variable band solver is actually the row-wise implementation of LU decomposition method where the bandwidth of the rows can vary from one row to another. Thus, in order to utilize this method, first the elements of the stiffness matrix must be stored by using the variable band storage scheme. In a variable band matrix storage scheme as shown in Figure 5.1, only the elements up to last non-zero element in each row are stored in a vector form. Similar to active column matrix storage, the location of the diagonal elements are stored in another vector, called 'RowId'. However, during LU decomposition, extra-storage space is required for the coefficients. For that reason, the row widths are arranged in such a way that the location of the last column of the $i+1^{th}$ row can not be before the last column of the ith row as shown in Figure 5.1.

The row-wise LU factorization is done by the application of the equations shown below (Equations 5.1-5.4). They are derived assuming variable band matrix storage.

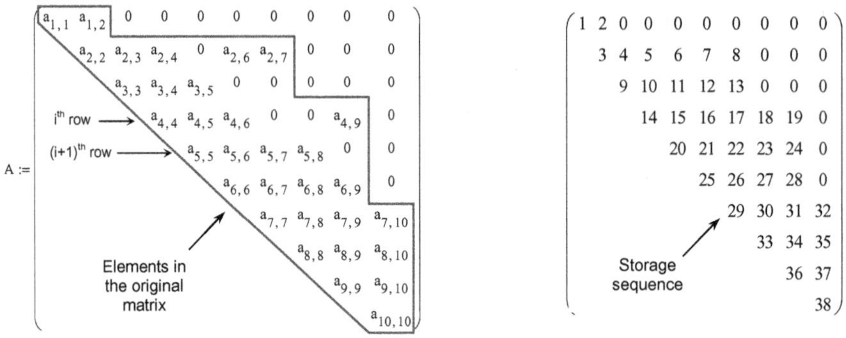

$$A := \begin{pmatrix}
a_{1,1} & a_{1,2} & 0 & 0 & 0 & 0 & 0 & 0 & 0 & 0 \\
& a_{2,2} & a_{2,3} & a_{2,4} & 0 & a_{2,6} & a_{2,7} & 0 & 0 & 0 \\
& & a_{3,3} & a_{3,4} & a_{3,5} & 0 & 0 & 0 & 0 & 0 \\
& & & a_{4,4} & a_{4,5} & a_{4,6} & 0 & 0 & a_{4,9} & 0 \\
& & & & a_{5,5} & a_{5,6} & a_{5,7} & a_{5,8} & 0 & 0 \\
& & & & & a_{6,6} & a_{6,7} & a_{6,8} & a_{6,9} & 0 \\
& & & & & & a_{7,7} & a_{7,8} & a_{7,9} & a_{7,10} \\
& & & & & & & a_{8,8} & a_{8,9} & a_{8,10} \\
& & & & & & & & a_{9,9} & a_{9,10} \\
& & & & & & & & & a_{10,10}
\end{pmatrix}$$

i^{th} row ⟶ (row 4)

$(i+1)^{th}$ row ⟶ (row 5)

Elements in the original matrix

$$\begin{pmatrix}
1 & 2 & 0 & 0 & 0 & 0 & 0 & 0 & 0 & 0 \\
3 & 4 & 5 & 6 & 7 & 8 & 0 & 0 & 0 \\
9 & 10 & 11 & 12 & 13 & 0 & 0 & 0 \\
14 & 15 & 16 & 17 & 18 & 19 & 0 \\
20 & 21 & 22 & 23 & 24 & 0 \\
25 & 26 & 27 & 28 & 0 \\
29 & 30 & 31 & 32 \\
33 & 34 & 35 \\
36 & 37 \\
38
\end{pmatrix}$$

Storage sequence

Rowld = [1 3 9 14 20 25 29 33 36 38] : Pointer array of location of diagonals

A = [$a_{1,1}$ $a_{1,2}$ $a_{2,2}$ $a_{2,3}$ $a_{2,4}$ $a_{2,6}$ $a_{2,7}$ ...] : Elements of the original matrix stored in vector form

Figure 5.1 Variable Band Storage Scheme

Let A be the positive definite, symmetric stiffness matrix and F be the load vector. The factorization starts with the calculation of the lower triangular coefficients of the pivot-row. The coefficients are calculated by dividing the upper diagonal coefficients with their diagonal as shown in Equation (5.1). By using these coefficients, the rest of the rows, non-pivot rows, are updated by using Equation (5.2):

$$L_{ij} = U_{ij} / D_{ii} \qquad\qquad i=1 \text{ to } BEQ\text{-}1 \;\; j=i+1 \text{ to } lz \qquad\qquad (5.1)$$

$$U_{jk} = A_{jk} - L_{ij} \cdot U_{ik} \qquad\qquad j=i+1 \text{ to } lz \quad k=j \text{ to } lz \qquad\qquad (5.2)$$

where BEQ = total number of equations at the interface
lz = minimum of either i+iz or BEQ
iz = row length of ith row
jz = maximum of either 1 or i-iz

After the triangular decomposition of the symmetric A matrix, the forward substitution is performed for each load case. Each F vector is updated by using Equation (5.3):

$$F_j = F_j - L_{ij} \cdot F_i \qquad\qquad j{=}1 \text{ to } BEQ \quad i{=}j \text{ to } lz \qquad (5.3)$$

Then, the system is solved by back substitution for each F vector using Equation (5.4):

$$F_j = F_j - U_{ji} \cdot F_i / D_{ii} \qquad\qquad i{=}BEQ \text{ to } 2 \quad j{=}i \text{ to } jz \qquad (5.4)$$

5.2.2 Operation Count for Serial Solution

The operation count for the row-wise LU decomposition algorithm is calculated for the factorization, forward substitution, and back substitution. These results are for the serial solution.

The factorization step is examined in two parts: pivot row operations, as shown in Equation (5.1), where the lower triangular coefficients of the pivot row are calculated and non-pivot row operations where the remaining rows are updated by using Equation (5.2). Assuming that the matrix has an average bandwidth of 'm' and number of equations of 'N', the operation count for factorization step is calculated as follows:

For each pivot row, there are 'm-1' divisions. Thus, the total number of operations to calculate lower triangular coefficients of pivot row (T_p) is:

$$T_p = \sum_1^{N-m}(m-1) + \sum_{i=N-m+1}^{N-1}(N-i) \qquad (5.5)$$

$$T_p = N \cdot m - \frac{m^2}{2} - N + \frac{m}{2}$$
(5.6)

Each pivot row updates 'm' non-pivot rows. The updates consist of 'm-j' multiplications and 'm-j' subtractions for each row where 'j' is the difference between the row numbers of the non-pivot and the pivot row. There are 'N-1' pivot row updates, but after $(N-m)^{th}$ row, the size of the bandwidth decreases. Thus, the total number of operations for non-pivot operations (T_{np}) is equal to:

$$T_{np} = \sum_{i=1}^{N-m} \sum_{j=1}^{m-1} 2 \cdot (m - j) + \sum_{i=N-m+1}^{N-1} \sum_{j=1}^{N-i} 2 \cdot (N - j - i)$$
(5.7)

$$T_{np} = N \cdot m^2 - \frac{2}{3} \cdot m^3 - N \cdot m + \frac{2}{3} \cdot m$$
(5.8)

The total number of operations for the factorization (T_{fact}) is the sum of Equation (5.6) and Equation (5.8) and equal to:

$$T_{fact} = N \cdot m^2 - \frac{2}{3} \cdot m^3 - \frac{1}{2} \cdot m^2 - N + \frac{7}{6} \cdot m$$
(5.9)

$$T_{fact} = O(m^2 N - 2/3m^3)$$
(5.10)

Forward substitution is composed of 'm' multiplications and 'm' subtractions for each row of the stiffness matrix. For 'b' load cases, the total number of operations for forward substitution (T_{fs}) is equal to:

$$T_{fs} = b \cdot \left\{ \sum_{i=1}^{N-m} 2 \cdot (m-1) + \sum_{i=N-m+1}^{N-1} 2 \cdot (N-i) \right\}$$

(5.11)

$$T_{fs} = b \cdot (2 \cdot N \cdot m - m^2 - 2 \cdot N + m)$$

(5.12)

The final step is back substitution. During back substitution, the ith displacement is calculated by dividing the ith row of the vector with the ith diagonal of the stiffness matrix. Then, the remaining of the load vector is modified by the upper triangular part of the ith row. In other words, there will be 'm' multiplications and 'm' subtractions for each row of the load vector. For b load vectors, the total number of operations for back substitution (T_{bs}) is:

$$T_{bs} = b \cdot \left\{ \sum_{i=1}^{N-m} 2 \cdot (m-1) + \sum_{i=N-m+1}^{N-1} 2 \cdot (N-i) + (N-1) \right\}$$

(5.13)

$$T_{bs} = b \cdot (2 \cdot N \cdot m - m^2 - N + m - 1)$$

(5.14)

Finally, the total number of operations for load factorization (T_{lf}) is calculated by adding Equation (5.12) with Equation (5.14) which is:

$$T_{lf} = b \cdot (4 \cdot N \cdot m - 2 \cdot m^2 - 3 \cdot N + 2 \cdot m - 1)$$

(5.15)

$$T_{lf} = 4 \cdot b \cdot O(N \cdot m - \frac{m^2}{2})$$

(5.16)

167

The above calculations indicate that the factorization time is mainly affected by the size of the bandwidth. Moreover, if the number of load cases is close to the size of the bandwidth, the time spent for the load factorization may be close to the time spent during the factorization of the interface stiffness.

5.2.3 Parallel Implementation

The pseudo-code shown in Figures 5.2 to 5.4 shows the parallel implementation of the row-wise LU decomposition method. For illustration purposes, the pseudo code is designed for a full matrix which has 'N' equations. It is assumed that the interface stiffness matrix was assembled and its rows were distributed to other processors in a cyclic manner as shown in Figure 5.2. In the pseudo code, the row numbers that are stored by a particular processor are kept in an array called 'map'. The jth value of the 'map' array shows whether the processor keeps the jth row of the stiffness matrix. If so, the jth value of the 'map' array is larger than zero. Each processor has positive values at the different locations of their 'map' arrays.

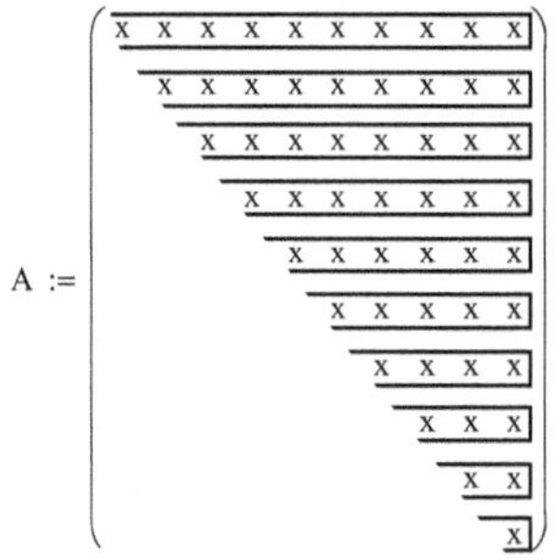

| | Proc. Id | 'Map' array | | |
		Processor 1	Processor 2	Processor 3
X X X X X X X X X X	... 1	1	0	0
X X X X X X X X X	... 2	0	1	0
X X X X X X X X	... 3	0	0	1
X X X X X X X	... 1	1	0	0
X X X X X X	... 2	0	1	0
X X X X X	... 3	0	0	1
X X X X	... 1	1	0	0
X X X	... 2	0	1	0
X X	... 3	0	0	1
X	... 1	1	0	0

$A :=$

Figure 5.2 Cyclic Data Distribution with 3 processors

The pseudo code for factorization is given in Figure 5.3. The algorithm starts with a 'k' loop where 'k' represents the current pivot row. The processor which retains the kth row, called the master processor, distributes the elements of this row to other processors. Then, each processor will calculate 'L' of that row by using Equation (5.1), but only the master processor stores it permanently. After having calculated the 'L', each processor updates their non-pivot rows according to Equation (5.2).

```
for k=1 to N-1 {
   Broadcast(kth row) // N-k+1 words

   // pivot column operations
   for i=k+1 to N {
      Lk,i=Mk,i/Mk,k
   }

   // non-pivot column operations
   for j=k+1 to N {
      if (map(j)>0) {
         for ii=j to N
            Mj,ii= Mj,ii-Lk,ii.Mk,j
      }
   }
}
```

Figure 5.3 Pseudo Code for Factorization

For the load factorization step, the algorithm assumes that the rows of the load vectors are distributed among the processors in the same way that the stiffness matrix is distributed. As shown in Figure 5.4, the processor that stores the kth row, first collects the elements of the kth row of all the load vectors in a vector called 'c'. Then, it distributes the vector 'c' to the other processors. After that, each processor modifies its load vector by using the kth row of the load vectors and i^{th} row of the stiffness matrix. The ii^{th} value of the 'map' array will be larger than zero if a processor retains the ii^{th} row.

169

```
for k=1 to N {
   if (map(k)>0) {
      for j=1 to b { // number of load cases
         c_j=F_k,j
      }
   }

   Broadcast(c) // b words
   for ii=k+1 to N {
      if (map(ii)>0) {
         for j=1 to b
            F_ii,j=F_ii,j-c_j.L_k,i
      }
   }
}
```

Figure 5.4 Pseudo Code for Forward Substitution

A very similar algorithm is used for the back substitution. This time, the kth rows of the load vectors are first divided by the kth diagonal of the stiffness matrix and then distributed to other processors. The pseudo code for back substitution is given in Figure 5.5.

```
for k=N to 1 {
   if (map(k)>0) {
      for j=1 to b { // number of load cases
         F_k,j=F_k,j/M_k,k
         c_j=F_k,j
      }
   }

   Broadcast(c) // b words
   for ii=k-1 to 1 {
      if (map(ii)>0) {
         for j=1 to b
            F_ii,j=F_ii,j-c_j.M_i,k
      }
   }
}
```

Figure 5.5 Pseudo Code for Back Substitution

5.2.4 Operation Count for Parallel Solution

The number of operations for the parallel implementation of the row-wise LU decomposition method is calculated using the following assumptions:

- Number of equations, N, is a factor of number of processors, p

- Each processor stores N/p rows of the stiffness matrix

- The rows are distributed in a cyclic manner

- Each processor has the same computation speed

The factorization consists of two steps, pivot row and non-pivot row operations. In pivot-row operations, there are 'm' divisions which are repeated 'N-1' times. This is the serial part of the algorithm. Therefore, the number of operations for pivot-row computations (T_p') is equal to:

$$T_p' = \sum_{1}^{N-m}(m-1) + \sum_{i=N-m+1}^{N-1}(N-i) \qquad (5.17)$$

$$T_p' = N \cdot m - \frac{m^2}{2} - N + \frac{m}{2} \qquad (5.18)$$

Each processor must wait until the lower triangular coefficients of the pivot row have been calculated. After that, the processors update the 'm/p' rows of the stiffness matrix. The updates consist of 'm' multiplications and subtractions. Since there are 'N-1' pivot row operations, the total number of operations for the non-pivot row updates (T_{np}') is equal to:

171

$$T_{np}{}' = \frac{1}{p} \cdot \left\{ \sum_{i=1}^{N-m} \sum_{j=1}^{m-1} 2 \cdot (m-j) + \sum_{i=N-m+1}^{N-1} \sum_{j=1}^{N-i} 2 \cdot (N-j-i) \right\} \qquad (5.19)$$

$$T_{np}{}' = \frac{1}{p} \cdot \left\{ N \cdot m^2 - \frac{2}{3} \cdot m^3 - N \cdot m + \frac{2}{3} \cdot m \right\} \qquad (5.20)$$

The total number of operations for the factorization ($T_{fact}{}'$) is calculated by adding Equation (5.18) and Equation (5.20). The result is given in the Equation (5.21) and its order is given in the Equation (5.22):

$$T_{fact}{}' = \frac{1}{p} \cdot \left\{ N \cdot m^2 - \frac{2}{3} \cdot m^3 - N \cdot m + \frac{2}{3} \cdot m \right\} + N \cdot m - \frac{m^2}{2} - N + \frac{m}{2} \qquad (5.21)$$

$$T_{fact}{}' = O\left(\frac{N \cdot m^2 - 2/3 \cdot m^3}{p} \right) \qquad (5.22)$$

The forward and back substitution steps use similar algorithms. Both modify 'm/p' rows of a single load vector. In back substitution, there will be an additional division for each row. For 'b' load cases, the number of operations for the forward ($T_{fs}{}'$) and the back substitution ($T_{bs}{}'$) are given in equations (5.24) and (5.26), respectively.

$$T_{fs}{}' = \frac{b}{p} \cdot \left\{ \sum_{i=1}^{N-m} 2 \cdot (m-1) + \sum_{i=N-m+1}^{N-1} 2 \cdot (N-i) \right\} \qquad (5.23)$$

$$T_{fs}{}' = \frac{b}{p} \cdot (2 \cdot N \cdot m - m^2 - 2 \cdot N + m) \qquad (5.24)$$

$$T_{bs}' = b \cdot \left\{ \frac{\displaystyle\sum_{i=1}^{N-m} 2 \cdot (m-1) + \sum_{i=N-m+1}^{N-1} 2 \cdot (N-i)}{p} + (N-1) \right\} \tag{5.25}$$

$$T_{bs}' = b \cdot \left(\frac{(2 \cdot N \cdot m - m^2 - 2 \cdot N + m)}{p} + (N-1) \right) \tag{5.26}$$

The total number of operations for the load factorizations (T_{lf}') is the summation of Equation (5.24) and Equation (5.26) which is equal to:

$$T_{lf}' = b \cdot \left\{ \frac{2 \cdot (2 \cdot N \cdot m - m^2 - 2 \cdot N + m)}{p} + (N-1) \right\} \tag{5.27}$$

$$T_{lf}' = 2 \cdot b \cdot O(\frac{2 \cdot m \cdot N - m^2}{p}) \tag{5.28}$$

As can be seen from Equations (5.22) and (5.28), the parallel version of the algorithm decreased the governing operations by a factor of 'p' for both factorization and load factorization. The only serial portion of the algorithm is the computation of the pivot row coefficients whose total number of operations is less than the non-pivot operations by a factor of '2m'. The algorithm seems to be very scalable in terms of computations, but there is also the communication cost between the processors.

5.2.5 Communication Cost

The communication time calculations model the amount of data that must be distributed and the amount of time required to perform this task for a given parallel algorithm. In such calculations, usually the linear communication time model [94] is utilized where the communication cost is represented by two variables, the start-up latency 'τ' and the data transfer time per word 'μ'. The start-up latency represents the amount of time required for initialization and data packing before sending the data. It mainly depends on the operating system and the parallel software library. The second variable, data transfer time per word, describes the amount of time required to transfer one word of data from one processor to another. It mainly depends on the communication hardware. It is possible to obtain these values for parallel machines like Intel-Delta or Cray T3D [107] from vendors or previous research, but for a PC Cluster, they must be computed.

The MPICH [115] library used in this study provides two basic send and receive methods each having four different data transfer modes for communication between processors:

- **Blocking send and receive:** The send method does not return until the message data is safely copied into the matching receive buffer or a temporary system buffer.

  o Standard mode: In this mode, MPI decides whether or not to buffer the outgoing messages.

  o Buffered mode: The send method returns immediately after the message is copied to the temporary buffer even if a matching receive has not been posted.

o Synchronous mode: The send method starts whether a matching receive has been posted or not, however it will not return unless the receive operation has been initiated.

o Ready mode: The send method can only be started if the matching receive is posted. Otherwise, the send operation is erroneous.

- **Non-blocking send and receive:** The non-blocking send call initiates the send operation and returns before the message was copied to the send buffer. Similarly, the non-blocking receive call initiates the receive operation and returns before the message is stored in the receive buffer. Similar to the blocking send and receive it has four different modes: standard, buffered, synchronous and ready.

These routines are utilized during point to point communication. If the data needs to be transferred to all processors in the cluster, a 'broadcast' routine is used. The 'broadcast' routine first checks the available parallel environment and the size of the data and then utilizes the most optimum send and receive routine for data transfers.

During the interface solution, most of the data communication occurs during the factorization step. Each pivot row must be transferred to all processors in order to calculate the lower triangular coefficients. Thus, there are 'N-1' transfers and each transfer consists of 'm' double data. The 'broadcast' routine was utilized for these transfers. Hence, while calculating the communication time, the total amount of data is multiplied by the broadcasting speed, 'α_p', of the parallel environment for a given number of processors, 'p'. The communication time for these transfers is calculated as shown in Equation (5.29):

$$CT_{low} = \alpha_p \cdot \left(\sum_{i=1}^{N-m} m + \sum_{i=N-m+1}^{N-1} (N-i+1) \right) \tag{5.29}$$

$$CT_{low} = \alpha_p \cdot \left[m \cdot (N - \frac{m-1}{2}) - 1 \right] \tag{5.30}$$

The 'broadcast' routine was used to transfer data among processors during forward and back substitution, however when the number of load cases is small (10-20), the broadcast routine performed blocking send and receive. For this reason, the communication cost during load factorization was computed by using the linear communication time model [94]. During the forward and back substitution, the same amount of data is communicated between processors. For both cases, there are 'N-1' transfers and each transfer carries 'b' double data. Thus, the communication time for both the forward and the back substitution (CT_s) is equal to:

$$CT_s = \tau \cdot (p-1) \cdot \sum_1^{N-1} 1 + \mu \cdot (p-1) \cdot \sum_1^{N-1} b \tag{5.31}$$

$$CT_s = \mu \cdot (p-1) \cdot b \cdot N + \tau \cdot (p-1) \cdot N - \mu \cdot (p-1) \cdot b - \tau \cdot (p-1) \tag{5.32}$$

The total communication time for the load factorization (CT_{lf}) is then equal to:

$$CT_{lf} = 2 \cdot CT_s \tag{5.33}$$

$$CT_{lf} = 2 \cdot \mu \cdot (p-1) \cdot b \cdot N + 2 \cdot \tau \cdot (p-1) \cdot N - 2 \cdot \mu \cdot (p-1) \cdot b - 2 \cdot \tau \cdot (p-1) \tag{5.34}$$

since the same communication is required for both the forward and back substitution.

5.3 Parallel Solution Time Estimations

5.3.1 Computation Time

The computational performance of a computer depends on many variables. In terms of hardware components, the processor speed, the memory speed, the size and the speed of the cache memory and external bus speed can be considered as key role players in the determination of the computational speed. In addition, the operating system, the programming language and the algorithm implementation are the software side elements which influence the performance. As a result, predicting the speed of computers by just looking at the properties of the hardware components is not possible because there are many variables that affect the performance of a computer. As a result, test runs are required to determine the computational properties of a computer.

There are also some concerns when using the test run approach. First of all, is it possible to assume that the computer performs all the multiplications at the same speed? For example, is there a difference between multiplying a scalar with a vector or a vector with another vector? Secondly, if two vectors are multiplied, do their sizes affect the speed of computation? In order to answer these questions and estimate the computation speed of the interface solution algorithm, a serial and a parallel version of the variable band solver were developed. They were tested with generated matrices having various sizes on different computers. For each run, the time spent during the factorization, the forward substitution and the back substitution was stored.

First, the computational properties of the computers belonging to three different clusters were tested by using the serial version of the variable band solver. The properties of these computers are given in Table 5.1. For each computer, the factorization, forward

177

and back substitution speeds were computed by dividing the operation count values with the time required to finalize the factorization, and the forward and back substitutions, respectively.

Table 5.1 Properties of the PC Clusters

Name	Processor	Memory	Cache Memory	Bus Speed	Operating System
DEC	Pentium 166 Mhz	64 Mb	?	66 Mhz	Windows NT
AFC	Celeron 400 Mhz	128 Mb	128 Kb	66 Mhz	Windows 2000
DELL	Pentium (4) 2.4 Ghz	1 Gb	512 Kb	800 Mhz	Windows XP

The governing operation of the serial variable band solver during factorization is the non-pivot row updates, Equation (5.2). The elements of the row 'j' are computed by subtracting the product of a scalar 'L_{ij}' with a vector, i.e. elements of row 'i'. Hence the non-pivot operations are actually a vector-scalar product followed by a vector-vector subtraction. Moreover, the elements of both vectors are stored sequentially in the memory.

The governing operation for forward substitution is very similar to the factorization, Equation (5.3). The scalar 'F_i' is multiplied with the vector 'L_{ij}' and subtracted from the vector 'F_j'. The elements of both vectors are also stored sequentially in the memory. On the other hand, the situation during the back substitution is different. Although the governing operations for the back substitution looks like a scalar-vector multiplication followed by a vector-vector subtraction as in Equation (5.4), the elements of the jth column of 'U' are not sequentially stored in the memory. In other words, each element is at a different memory location. For that reason, the number of data transfers from the memory to the cache memory increases and the performance of the computation decreases.

Table 5.2 shows the computational speeds of three different computers during factorization, forward and back substitution calculated by averaging the speed values obtained from the solution of generated matrices having a constant number of equations with different bandwidths for 100 load cases. The unit of the computation speed is represented in 'flops' (floating-point operations per second) for double precision operations.

Table 5.2 Computation Speeds for Serial Version

Name	Factorization	Forward Subs.	Back Subs.
DEC	11.75 Mflops *	12.14 Mflops *	2.83 Mflops *
AFC	27.17 Mflops	34.49 Mflops	22.99 Mflops *
DELL	363.36 Mflops	444.44 Mflops	128.21 Mflops *

Speed varies as bandwidth increases

The back substitution speeds are much slower than the forward substitution and the factorization speeds for all the computers tested. The computation speed of the DEC computer dropped as the size of the problem increased for the factorization, forward, and back substitution. Similar speed variations are observed during the back substitution for the AFC and the DELL computers. The speed decreased as the bandwidth increased due to the increased cost of non-sequential memory access.

The row-wise factorization speed of the AFC computer (27.17 Mflops) is slower than the column-wise factorization speed (38.85 Mflops, Figure 3.8). However, the DELL computer performed row-wise factorization (363.36 Mflops) 2.2 times faster than

179

column-wise factorization (166.58 Mflops). This indicates that a scalar-vector product is much faster than a vector-vector product on the DELL computer. For both AFC and DELL computers, the forward substitution speed is slightly faster than the factorization speed.

The parallel implementation of the variable band solver has a more complex structure than the serial version. The rows of the stiffness matrix and the elements of the force vectors are stored at the processors in a cyclic manner. Hence, none of the processors store two or more consecutive rows. As a result, the elements of 'L' are accessed non-sequentially on each processor during the forward substitution. Normally, this will result in a decrease in the computation speed for the forward substitution. However, for the multiple loading condition problems, the elements of 'L' that will be used to factorize the loads are copied to another vector in an ordered way and this vector is used during the forward substitution. As a result, the decrease in computational speed due to non-sequential memory access is avoided. A similar approach is utilized during back substitutions also.

The parallel version of the variable band solver was tested on generated matrices with a constant number of equations and a varying uniform bandwidth. The speeds for the factorization, the forward and the back substitution were computed separately for each processor. Table 5.3 shows the average computational speeds for the DEC, AFC and DELL computers using this parallel version. During these runs, each matrix was solved using two computers.

Table 5.3 Computation Speeds for Parallel Version

Name	Factorization	Forward Subs.	Back Subs.
DEC	11.75 Mflops *	9.43 Mflops *	12.06 Mflops *
AFC	27.17 Mflops	31.64 Mflops *	32.36 Mflops *
DELL	363.63 Mflops	336.70 Mflops *	344.83 Mflops *

Speed varies as bandwidth increases

As can be seen from Table 5.3, the factorization speeds remained the same in the parallel version of the algorithm. The forward substitution speeds were slightly slower than the serial version due to the cost of copying the elements of 'L' to another buffer in an ordered way. As the size of the bandwidth increased, the cost of the copying operation increased, thus the forward substitution speed decreased. On the other hand, the back substitution speed increased significantly. Using an ordered temporary buffer increased the back-substitution speed by 2.7 times on the DELL Cluster. Similar to the forward substitution, back substitution speed decreased as the bandwidth increased.

5.3.2 Communication Time

The communication speed of a PC cluster will depend on the brand and the type of the network switch, network cards, and the PC's hardware. With the huge variety of these components, it is very difficult to predict the communication time. In an effort to experimentally determine the communication performance, a series of test runs were performed on the three different computer clusters used in this study.

First, a computer program was developed in order to determine the data transfer time per word for a particular cluster. In this program, a series of data chunks were transferred from the master processor to the slaves by using blocking send and receive routine. In this routine, the master processor sends data to a single slave processor and waits until all

181

data has been sent. Then it starts to send data to the other processor. Thus, it is expected that the communication cost of the blocking send and receive increases linearly as the number of processors increases. For each data chunk, the transfer times for each processor were stored. Once all transfers had been completed, a linear regression analysis was performed between the size of the data chunks and their transfer times to each processor. The start-up latency cost was ignored during the regression analysis since its value was negligible when compared with the data transfer cost. Thus the trend line started from the origin. The slope of the trend line indicated the data transfer time per word.

First, these runs were performed on the DEC Cluster which had 8 computers with 166 Mhz Intel Pentium Processor, 64 MB memory and 66 Mhz external bus speed. All the computers were configured with 10Mbit network cards. They were connected using a Netgear FS108 dual speed (10/100Mbit) ethernet switch. All computers were identical, i.e. they were from the same manufacturer, they had the same motherboards etc. Furthermore, all computers had the same operating system, Windows NT 4.0.

Figure 5.6 presents the blocking send and receive timings on the DEC cluster with 6 computers. The behavior of this cluster was as expected. There was a linear relationship between the amount of data transferred and the transfer time. The time required to send data to the third computer was two times slower than the time for the second computer. The per data transfer time for this cluster was computed by first subtracting the slope of the third line from the second line, the fourth line from the third line and so on. Then the average of the results was considered as the data transfer time per word and was equal to 1.282×10^{-3} secs/KB. In other words 6.24 Mbits of data was transferred in one second.

182

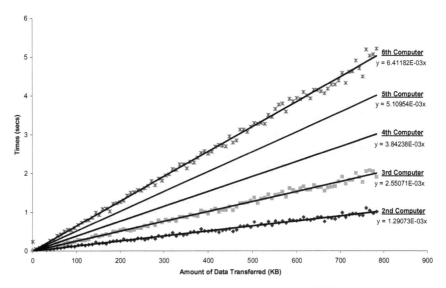

Figure 5.6 Blocking Send & Receive Results on DEC Cluster

Similar runs were repeated on the AFC Cluster. The AFC Cluster was composed of 8 computers each with 400 Mhz Intel Celeron processor, 128 MB RAM, 128 KB L2 Cache memory with 66 MHz external bus speed. These computers are commonly referred as clones. Each of them had different motherboards and network cards. They all had Windows 2000 operation system. The network cards had a speed of 100 Mbits and were connected using a Netgear FS108 dual speed (10/100 Mbit) ethernet switch.

Figure 5.7 shows the results for the AFC Cluster. Although the data transfers were between two computers, there were significant slow downs at random locations. Still, the behavior can be considered as linear. The per data transfer speed for this cluster was computed as 1.273×10^{-4} secs/KB or, in other words, 62.82 MBits of data was transferred in one second.

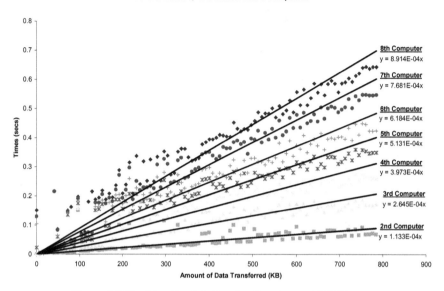

Figure 5.7 Blocking Send & Receive Results on AFC Cluster

The third cluster is the DELL Cluster. The DELL Cluster had 12 identical computers connected to the CEE department's network system. Each computer had Pentium IV 2.4 GHz processors, 1 Gbyte memory, 512 Kbyte cache memory and 800 Mhz bus speed. The network cards had a speed of 1 Gbits. Windows XP Professional was installed on each computer.

The communication times for blocking send & receive operations on the DELL Cluster are presented Figure 5.8. Similar to the other clusters, there was a linear relationship between the communication time and the amount of data transferred. The data transfer time per word was found to be 0.856×10^{-5} secs/Kbyte or 91.24 Mbits per second.

184

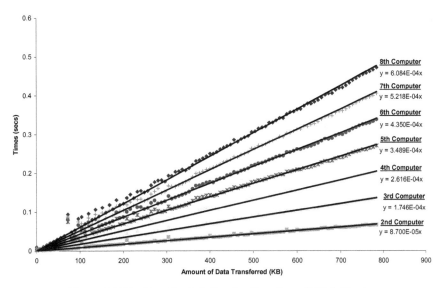

Figure 5.8 Blocking Send & Receive Results on DELL Cluster

A procedure similar to blocking send and receive speed determination was applied to compute the broadcasting speeds of clusters. This time, the data chunks were distributed to the processors by using the 'broadcast' routine. Each processor stored the timings after they complete receiving the data. At the end, a linear regression analysis was performed for each processor. The slope of the trend line then yielded the broadcast speed for that processor. It was assumed that the slowest broadcasting speed governed the communication time and was considered to be the broadcast speed of the cluster for the given number of processors.

Figure 5.9 shows the broadcast times for DEC cluster for 2, 4, 6, and 8 computers, respectively. Similar to the send and receive results, a linear behavior was observed. The broadcast speed for two computers was very close to the send and receive speed whereas

185

the broadcast speeds for 4, 6 and 8 computers were faster than the corresponding send and receive speeds. When the data was distributed among 6 computers, two computers received the data faster than the other three. On the other hand, all the computers received the data at the same time when the data was broadcasted among 8 computers. Figure 5.9 also presents the broadcasting speeds of the DEC Cluster. The slowest computer was considered for the speed determination.

The broadcast times of AFC cluster with 2 to 8 computers is shown in Figure 5.10. The results in AFC Cluster were not as uniform as those from the DEC Cluster. The data was spread on a triangular area rather than following a linear path as shown in Figure 5.10. Moreover, the broadcast speeds obtained from the regression analysis were much slower than the blocking send and receive speeds for two and four computers.

The broadcasting speeds for the DELL cluster are shown in Figure 5.11. Similar to the results of AFC cluster, the data did not follow a linear path when the size of data was more than 300 Kbytes. Moreover, the broadcast speed obtained from the linear regression analysis for two and four computers was slower than the blocking send and receive speeds.

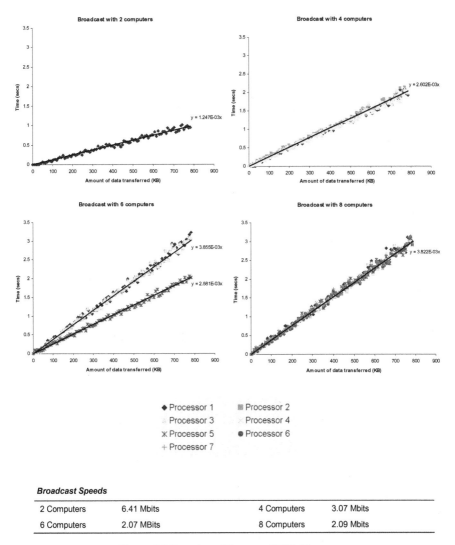

Figure 5.9 Broadcast Times on DEC Cluster

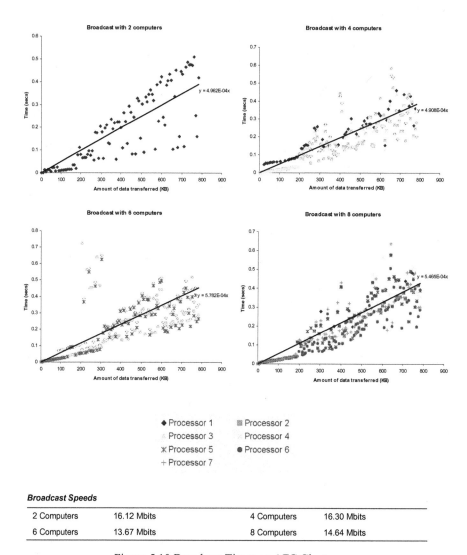

Figure 5.10 Broadcast Times on AFC Cluster

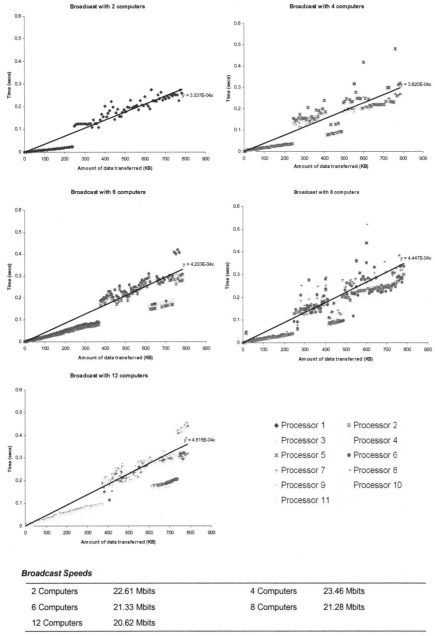

Figure 5.11 Broadcast Times on DELL Cluster

Another series of runs were performed using the actual models in Appendix A in order to compute the broadcast speeds of the DELL cluster. The results were compared with the results obtained from the regression analysis. The broadcast speeds were monitored during the factorization of the interface stiffness matrix of the example problems solved in this study (Appendix A). At the end of each factorization, the broadcast speeds were obtained by dividing the total amount of data that was transferred with the total time spend during broadcast operation. Table 5.4 presents the slowest, the fastest and the average broadcast speeds for 2 to 8 processor solutions on the DELL cluster.

Table 5.4 Broadcast Speeds of DELL Cluster

# Procs.	Slowest (Mbits)	Fastest (Mbits)	Average(Mbits)
2	46.24	64.00	54.05
4	32.79	40.40	35.71
6	19.66	26.58	22.47
8	22.54	27.03	24.54

These results indicated that the broadcast speed varied considerably. For these example problems, the differences in speeds for the same number of processors were around 35%. Moreover, the average broadcast speed between 2 processors was slower than the blocking send and receive speed but as the number of processors increased, the broadcast operation became faster. The results obtained from the regression analysis overestimated the speeds for 2 and 4 processors but found similar speed values for 6 and 8 processors.

5.3.3 Performance Analysis

The computation and the communication speeds obtained in the previous sections were utilized to estimate the parallel solution time of generated linear problems with matrices having constant bandwidths. Then, the estimated times were compared with the actual solution times to see if the estimations using the computation and communication speeds were able to predict the solution time within an acceptable error range.

The solution time estimations composed of two parts: stiffness factorization and load factorizations. Developing a time estimate for the load factorization presented difficulties. First of all, the forward and back substitution speeds varied as the size of the bandwidth changed. Moreover, when the number of loading conditions was small (<100), the communication cost during load factorization highly depended on the start-up latency value which could not be successfully computed. On the other hand, the stiffness factorization time was approximately 'm/4·b' times slower than the load factorization time (Equation (5.22) / Equation (5.28)) where 'm' is the average bandwidth and 'b' is the number of loading conditions. For dense matrices, the parallel solution time was mostly governed by the stiffness factorization step. For that reason, the solution time estimations considered only the factorization step.

The factorization time was estimated by adding the computation time and the communication time. The time spent during the computations was computed by dividing the operation count values obtained from Equation (5.21) with the corresponding speed value from Table 5.3. Similarly, the communication time was computed by dividing the total amount of data that was transferred, Equation (5.30), with the broadcast speeds for the given number of processors. For the AFC cluster, the broadcast speeds for 2 and 4

191

processors were taken as equal to the blocking send and receive speeds. For the DELL cluster, the average broadcast speeds from Table 5.5 were utilized.

First, the DEC cluster was tested to see if the estimated solution times compared favorably with the actual results. The solution was performed on generated matrices having uniform bandwidths. The size of the matrix was chosen in such a way that, when the full matrix was distributed to two processors, the required storage space did not exceed the available memory. Figure 5.12 shows the results with 2, 4, 6 and 8 processors. All the matrices had 3000 equations but the bandwidth ranged between 100 to 3000.

There was a good agreement with the analytical and the experimental results for 2, 4 and 8 processor solutions. The difference between the experimental and the analytical times are more pronounced for the 6 processor solution but the average percent difference for matrices having a bandwidth between 1000 and 3000 was 4%. For other cases, the average difference was less than 2%.

Figure 5.13 presents the results of the similar runs on AFC cluster. The generated matrices had 5000 equations with uniform bandwidths ranging from 100 to 5000. The analytical and the experimental results were close for 2 to 8 processor solutions. The average difference is less than 10% for all cases.

Figure 5.14 presents the results of the similar runs on DELL cluster. The generated matrices had 10000 equations with uniform bandwidths ranging from 1000 to 10000. The error in the time estimations increased as the number of processors increased. The error in the factorization times with two processors was around 6% whereas the average error increased to 12.5% when 12 processors were used. This is mainly due to the increased effect of variable broadcasting speeds as discussed in Section 5.3.3.

192

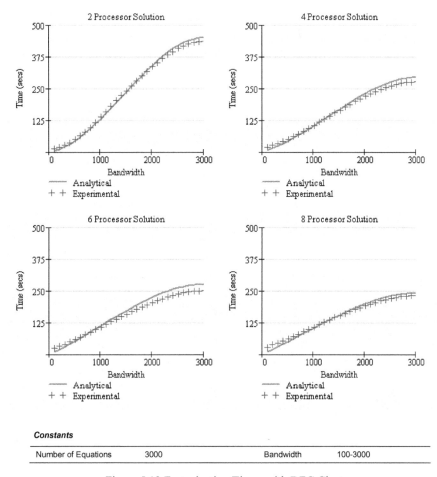

Constants

Number of Equations	3000		Bandwidth	100-3000

Figure 5.12 Factorization Times with DEC Cluster

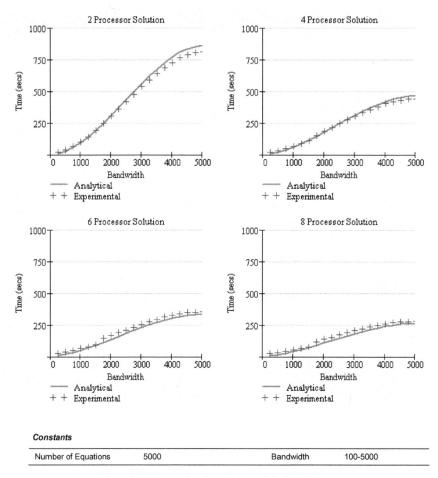

Figure 5.13 Factorization Times with AFC Cluster

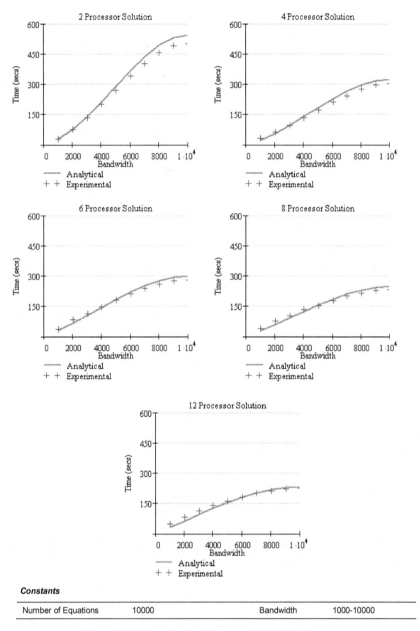

Figure 5.14 Factorization Times with DELL Cluster

5.3.4 Interface Solution of Actual Models

In general, the time estimations gave close results to the actual times for the parallel solutions of matrices having constant bandwidth (average error <10%). However, the actual problems will have variable bandwidths. Thus, the same estimation technique was applied to the interface solutions of the example problems and the estimations were compared with the actual results. The operation count values were computed numerically.

The following Tables 5.5 and 5.6 show the time estimations, the real solution times, and the percentage differences for the interface solution of the 2D Square Mesh (Appendix A.1.1) and the Nuclear Waste Plant (Appendix A.1.6) models on the DELL cluster using 2 to 12 processors.

Table 5.5 Interface Factorization Times of 2D Square Mesh

Number of Processors	Time Estimations	Actual Factorization Time	% Difference
2	1.44 secs	1.33 secs	+8.2%
4	5.05 secs	5.19 secs	-2.7%
6	15.35 secs	16.67 secs	-7.9%
8	20.17 secs	19.25 secs	+4.8%
12	34.87 secs	40.62 secs	-14.1%

Table 5.6 Interface Factorization Times of Nuclear Waste Plant

Number of Processors	Time Estimations	Actual Factorization Time	% Difference
2	10.06 secs	8.58 secs	+17.2%
4	84.93 secs	92.34 secs	-8.0%
6	201.95 secs	206.46 secs	-2.2%
8	166.53 secs	143.31 secs	+16.2%
12	414.76 secs	376.01 secs	+10.3%

The time estimations were in a good agreement with the actual factorization times. The maximum error occurred during the interface solution of the Nuclear Waste Plant model with 2 processors. This is due to the overestimation of the broadcast speed for 2 processors. The time estimation for the 12 processor solution of the 2D Square Mesh model underestimated the solution time whereas the estimations for Nuclear Waste Plant model overestimated the solution time by 10.3%. Again, this is mainly due to the varying broadcast speed. Overall, the error was less than 20%, and based on these results, one can conclude that it is possible to predict the parallel factorization time within 20%.

5.3.5 Performance Comparison

Up to this point, the communication and the computation performances of the three clusters were examined. The results were tested by comparing the solution time estimation for the parallel variable band solver with the actual results obtained from the parallel solutions. The estimations were able to catch the actual results with a maximum error of 20% which showed that both communication and computation speed values computed for each cluster were able to represent their parallel behavior.

Table 5.7 presents the factorization speed, the blocking send and receive speed and the broadcast speed with 8 processors. The last column of the table shows the ratio of the broadcast speed and the factorization speed (communication/computation).

Table 5.7 Computation and Communication Speeds of all Clusters

Cluster	Factorization	Blocking S&R	Broadcast 8	Communication / Computation
DEC	11.75 Mflops	6.24 Mbits	2.09 Mbits	0.18
AFC	27.17 Mflops	62.83 Mbits	14.65 Mbits	0.54
DELL	363.36 Mflops	91.24 Mbits	24.54 Mbits	0.07

The computers belonging to DEC and AFC clusters have a bus speed of 66 Mhz which is very slow when compared with today's typical PC. Their factorization time ratios are close to their processor speed ratios. The factorization speed of the AFC computer is 2.3 times faster than the DEC computer whereas the processor speed of the AFC computer is 2.4 times faster. On the other hand, the processor speed of the DELL computer is 6 times faster than the AFC computer but the factorization speed is 13 times faster. This may be due to the faster bus speed of the DELL computer.

There was a factor of 10 difference between the DEC and the AFC clusters in their blocking send and receive speeds which is the same difference between their network card speeds. The communication speed of the DELL cluster was less than expected due to the network hub and cables. Although its computers had 1Gbit network cards, its blocking send and receive speed was determined to be only 1.5 times faster than the blocking send and receive speed of the AFC cluster.

Figure 5.15 shows the speed-up values of the three clusters based on analytical results as the number of processors were increased for matrices having 5,000 equations with a bandwidth ranging from 100-5000. In all clusters, the speed-up increased as the bandwidth increased. The best speed-up was obtained using the AFC cluster, the parallel solution was nearly 5 times faster than the serial one with 8 processors. On the other hand, the DELL cluster performed the worst. The maximum speed-up was equal to 2 using 8 processors. The reason of such a drop in the speed-up performance lies in the communication/computation speed ratio. When this ratio in a cluster is larger, the parallel variable band solver is more scalable, which is the situation at AFC Cluster. The DELL cluster has the lowest communication/computation speed ratio, thus has the worst efficiency.

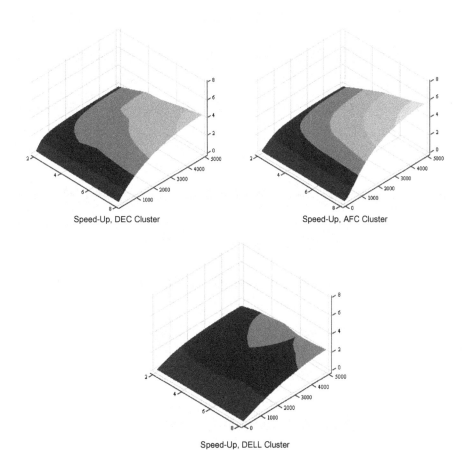

Figure 5.15 Speed-up values at DEC, AFC and DELL clusters

Tables 5.8 and 5.9 present the actual times spent for computation and communication during the parallel factorization of the interface stiffness matrices of the 2D Square Mesh (Appendix A.1.1) and the Nuclear Waste Plant (Appendix A.1.6) models, respectively at DELL Cluster. Most of the solution time was spent for the communication between processors. For example when 12 processors were utilized to factorize the interface stiffness matrix of the 2D Square Mesh model, only 10% of the total factorization time was spent for computation. On the other hand, this percentage was increased for the Nuclear Waste Plant model, computation time was 32% of the total factorization time due to the model's interface stiffness matrix having a larger bandwidth when compared with the 2D Square Mesh model. For both models, the ratio of the computation time to the solution time decreased as the number of processors increased. Even though the processors finished their assigned computations very rapidly, the communication speed of the cluster was so slow that the interface solution time was governed by the communication cost. That's why the parallel variable solver was not very scalable for this cluster.

Table 5.8 Detailed Factorization Times of 2D Square Mesh at DELL Cluster

Number of Processors	Computation Time	Communication Time	Factorization Time
2	0.49	0.84	1.33 secs
4	1.42	3.77	5.19 secs
6	2.75	13.92	16.67 secs
8	3.23	16.02	19.25 secs
12	4.24	36.38	40.62 secs

Table 5.9 Detailed Factorization Times of Nuclear Waste Plant at DELL Cluster

Number of Processors	Computation Time	Communication Time	Factorization Time
2	5.78	2.80	8.58 secs
4	46.78	45.56	92.34 secs
6	91.51	114.95	206.46 secs
8	54.29	89.02	143.31 secs
12	121.05	254.96	376.01 secs

5.4 Discussion of Results and Conclusions

In this chapter, the parallel variable band solver utilized for the interface solution of the substructure based solution framework was presented. Moreover, the computation and the communication properties of three different clusters were examined. Finally, the analytical solution time estimations were calculated for the solution of the interface problem of the example problems.

The variable band solver decreases the communication requirement between the processors considerably. For problems with multiple loading conditions, the lower triangular coefficients of each row are transferred to other processors once and utilized for both factorization and forward substitutions. During the back substitution, the total amount of communication is equal to the size of the load vector.

The computation speed of a computer highly depends on the way the memory is utilized. For example, the factorization and the back substitution speeds of the serial variable band solver were different although both performed scalar-vector products. The back substitution speed was slower due to non-sequential memory access. Moreover, the row-wise factorization speed was not equal to the column-wise factorization speed because the governing operation for the row-wise version was scalar-vector products whereas the column-wise version was governed by the vector-vector products. The DELL computer performed row-wise factorizations (363.36 Mflops) 2.2 times faster than the column-wise factorizations (166.58 Mflops). On the other hand, the row-wise factorization speed (27.17 Mflops) was slower than the column-wise factorization speed (38.85 Mflops) for the AFC computer. Thus, the computation speeds vary significantly depending on the type of operation and the computer hardware. Therefore, the computation speed must be separately computed for each operation for each computer.

The communication properties of the clusters were computed in two parts. First, the blocking send and receive performances were tested. In all clusters, there was a linear relationship between the amount of data transferred and the transfer time. In other words, the blocking send and receive speed was almost constant. The second part involved the computation of the broadcast speeds. The results in the DEC cluster were linear, however the other two clusters showed pretty irregular behavior. The broadcast speed varied with different amount of data being broadcast. Some variation was also observed during the interface solution of the Appendix A models. The variation in the broadcast speed of the DELL cluster was computed as approximately 35%.

The analytical time estimations were computed for the parallel variable band solver by using the computational and communication speeds of clusters. First, the time estimations were compared with the actual results for matrices having constant bandwidth. The average error remained below 10% at all clusters. When the interface problem of the example problems was solved by the parallel variable solver, the time estimations were able to predict the factorization time within 20%.

When the performances of three clusters were compared, it was observed that the scalability of the variable band solver with cyclic data distribution highly depended on the communication/computation speed ratio of the parallel environment. As this ratio decreased, the scalability decreased.

CHAPTER 6

IMPLEMENTATION

6.1 Introduction

Parallel solution techniques have been implemented in many finite element codes due to the increased availability of parallel computers including the use of PC clusters for parallel computations. Extensive research on parallel solution algorithms has been performed over the last twenty years. Past research has not only investigated the rewriting of the solution algorithms designed for sequential computing but has also developed entirely new strategies and algorithms to take advantage of the characteristics of various parallel computing architectures. Today, there are numerous parallel solution methods with iterative [62, 76] or direct solvers [55, 81, 82, 89]. The solution methods may be based on global, element-by-element or domain-by-domain strategies. However, their performance may be limited depending on the type of the analysis, the parallel environment, and the structural properties of the system.

Domain-by-domain methods generally perform the solution in two parts. The first part is called the local solution where the subdomain equations are constructed and transferred to the subdomain interfaces. Then, the interface solution, which is the second part of the solution method, starts where the interface displacements are computed. Farhat et al. [81] implemented the substructuring method, which is a type of domain-by-domain solution approach, for a parallel finite element program. They utilized direct methods for both the local and the interface solution. The interface solution algorithm was a row oriented formulation of LDL^T method. The proposed method had a generalized structure, in other words; it was designed to be utilized for not only the linear static but also the dynamic and the non-linear problems. The forward and back substitutions were performed after the stiffness matrix was factorized which increased the communication overhead. In their following work, Farhat and Wilson [79] developed an active column solver based on LU decomposition to perform the global stiffness matrix solution in parallel. In this method, the back-substitutions were performed serially. Both Farhat et al. [81] and Farhat and Wilson's [79] solution approaches were performed with high speed-up values up to 8 processors on parallel computers with distributed memory architecture.

In the following years, Farhat and Roux [76] derived a new version of an iterative domain decomposition method called the finite element tearing and interchanging (FETI) method. The method was based on removing the continuity constraint between the subdomains by using Lagrange multipliers and required the elimination of the rigid body modes of each subdomain. First, the rigid body modes were related to the Lagrange multipliers through an orthogonality condition. Then, the coupled system was solved by parallel conjugate gradient method. The proposed method was highly scalable because it decreased the communication overhead significantly. The comparative study carried out by Bitzarakis et al. [62] investigated three different iterative domain decomposition methods. The first method utilized preconditioned conjugate gradient method (PCG) on the global stiffness matrix, whereas the second method applied PCG method to the subdomain interfaces after eliminating the internal degrees of freedom of the subdomains. The third method was the implementation of FETI method. Among these three methods, FETI performed the fastest.

The FETI method is an iterative method which is not suitable for problems having multiple load cases because the solution must start from scratch for each load case. In order to overcome this deficiency of the FETI solver, Farhat et al. [76] developed a method that utilized the K-orthogonal subspace vectors from the previous loadings to decrease the number of iterations for the next loading. However, Bitzarakis et al. [62] showed that the performance of this approach decreased as the loading patterns vary.

The other deficiency of FETI method is the requirement that subdomains be free of mechanisms. In other words, the entire subdomain zero energy modes must be correctly predicted and eliminated before the solution. Day et al. [6] showed that FETI method failed when a 3D model contained mixed dimensional elements because when such models were partitioned, the subdomain had additional unpredicted mechanisms. They developed an "elastic connectivity graph" that avoided mechanisms for models having both 2D and 3D elements. For 1D elements, they recommended using multi-point constraints instead of using beam elements.

Fulton and Su [82] implemented the substructuring method on a shared memory computer. They utilized active-column storage scheme and a direct condensation algorithm. In order to balance the varying condensation times of substructures, they developed a method that assigned more processors to the substructures which were estimated to require more computation. In a following work, Synn and Fulton [100] investigated the following issues: direct or iterative solution, optimum number of processors for solution and global or substructure based solution. They recommended the direct solution approaches even though the iterative solvers were more scalable. Moreover, they derived operation count equations to estimate the optimum number of processors and to choose whether to use the global solution instead of substructuring.

Another substructuring based solution framework was developed by Hsieh et al. [89]. They utilized active column solvers for the interface solution but obtained low speed-up values. For this reason, they later utilized a serial direct solver for the interface problem [102]. Yang and Hsieh [102] also focused on the workload balancing problem of substructuring methods when direct condensation algorithms were utilized. They developed an iterative workload balancing algorithm that tried to balance the condensation time of substructures for sparse solvers. Their method workload balancing method consumed considerable amount of time and was found to be more suitable for dynamic or non-linear problems.

Multifrontal method is another direct solution method for distributed memory parallel architectures. It was derived from the frontal method, which was originally developed by Irons [90] where the factorization proceeds as a sequence of partial factorizations on a full submatrix called frontal matrices. Multifrontal methods perform computations on multiple independent fronts simultaneously. The first step of such methods is the construction of the multifrontal tree which is utilized to synchronize the parallel solution. Thus, the parallel efficiency and the memory usage of this method highly depends on the shape of the multifrontal tree. Guermouche et al. [83] performed a comparative study on various algorithms that developed multifrontal trees to determine how the multifrontal tree affected the memory requirement of the solution. They concluded that deep unbalanced trees were better than wide ones. The MUMPS code [55], which was a part of EU PARASOL Project [96], was an implementation of the parallel multifrontal method. After having computed the multifrontal tree, the MUMPS code [55] performed factorization, and forward and back substitutions. Based on the results of several test problems, it was concluded that the algorithm decreased the solution time when using up to 16 processors. Moreover, the method suffered from insufficient in-core memory for large problems and its efficiency significantly dropped due to page swapping.

Many structural engineering problems require linear analysis of large models which contain multiple loading cases. Iterative solvers are not well suited for such cases as they must start the solution from scratch for each loading case. Although there have been proposed methods to improve the efficiency of iterative solvers, their performances may decrease as the loadings vary. Moreover, the most efficient iterative method, FETI, is highly affected by the shape of the subdomains. If the subdomains are not mechanism free or all the rigid body modes of the unsupported subdomains are not predicted, the solution will not be obtained. Also, the method aborts for models having 1D elements.

The global solution approaches, such as multifrontal method, focus on speeding-up the solution of the linear systems only. Therefore, they require the entire assembled stiffness matrix. Moreover, the multifrontal methods work with multifrontal trees that affects the way the distribution of elements of the stiffness matrix are distributed among the processors. Thus, such methods necessitate very efficient implementation of stiffness and force matrix generation, their distribution, nodal displacement distribution, and element force computations. Moreover, global solvers must implement efficient out-of-core assembly, data-distribution, and solution algorithms for large structures.

The substructuring method, on the other hand, has the capability of parallelizing every step of the solution process, from element stiffness to element stress and force computations. If direct solvers are utilized, not only can they handle all shapes of the substructures but also become very suitable for problems with multiple load cases. In addition, out-of-core solvers can be easily implemented and efficiently utilized. On the other hand, current partitioning algorithms may create substructures with poorly balanced condensation times. This deficiency may significantly slow down the parallel solution. Furthermore, substructuring methods require very efficient interface solution algorithms for dense interface stiffness matrices.

This chapter focuses on the implementation of the substructure based parallel solution framework for large structural models with multiple loading cases. The solution starts by partitioning the structural model into substructures. Then, the imbalance in the estimated condensation times of substructures is decreased iteratively by transferring nodes among substructures. This way, one of the sources of inefficiencies of substructure base methods is improved. Next, the solution starts. The local solution is performed with a direct active column solver. The interface equations are assembled in parallel and solved with the parallel variable band solver. Finally, actual structural models obtained from AE's were tested to observe the efficiency of the framework.

6.2 Method

The solution framework currently consists of two separate programs that communicate via text files as shown in Figure 6.1. The first program is responsible from preparing the data for the parallel solution. It first reads the structural data, i.e. nodal coordinates, element connectivity, loadings etc. Then, it performs partitioning, workload balancing and equation numbering using the technique described in Chapter 4. Finally, it prepares the input file which contains the substructure definitions and the equation numberings for the parallel solution program.

The second program is a fully parallel finite element program. It is capable of performing element stiffness computations, assembly, solution and element force computations in parallel. It has an object-oriented database structure whose details are given in Appendix B. When the program completes the solution, it prepares the output for post-processing.

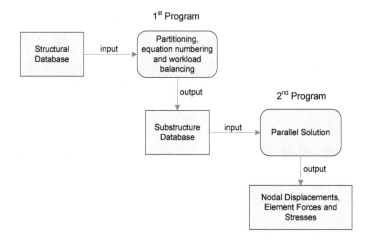

Figure 6.1 Solution Framework Structure

Both programs were developed with C++ and FORTRAN programming languages and utilized MPICH [115] message passing library for parallelization. They both work under the Windows operating system.

6.2.1 Data Preparation

The first step of a substructure based solution method is to divide the structure into substructures. This division is performed by using a partitioning algorithm. However, the current partitioning algorithms can not create substructures that have balanced condensation times when direct solvers are utilized. For this reason, a workload balancing step is added to the data preparation phase that iteratively adjusts the estimated imbalance of structures by transferring nodes from the substructures with slower estimated condensation times to the substructures with faster estimated condensation times.

The first program is developed for that purpose. It is not only responsible from partitioning but also workload balancing and equation numbering. First, the structure is partitioned into substructures by using the partitioning algorithms provided by METIS [30], a multilevel graph partitioning library. Next, the condensation times of each substructure are estimated by using the operation counts for condensation developed in Chapter 3 based on an active column scheme. Then, the substructures are modified in order to balance their condensation times by using PARMETIS [28], a parallel multilevel graph partitioning and repartitioning library. After that, the condensation times of the new substructures are estimated and the imbalance is resolved by repartitioning again. This procedure is repeated until either the condensation times of substructures are balanced or the maximum number of iterations has been reached.

211

The workload balancing approach assumes that balancing the condensation times will decrease the time spent during condensation. However, obtaining well-balanced substructures depends highly on the irregularity of the structure. For some structures, it is not possible to say a balanced solution exists for the given number of substructures. Moreover, there may be some cases where even the algorithm converges to a balanced solution, one of the partitioning created during the iterations produces less condensation times. For that reason, all the partitioning information obtained during each iteration is stored. Once the iterations end, the partitioning that has the fastest estimated solution time needs to be selected for the actual partitioning.

The solution time estimations can consider the condensation time only or the total solution time including the interface solution time estimation. In this study, it was observed that the balanced substructures had a larger interface problem than the initial substructures. Moreover, as the number of processors increased, the interface solution time began to govern the total solution. For these reasons, when the iterations are finalized, the total solution times (condensation + interface solution) are estimated for each partitioning and the fastest one is chosen for the solution.

In order to predict the total solution time, the parallel interface solution time must be estimated. One of the difficulties of interface solution time estimation is the determination of the number of interface equations and the bandwidth. In order to calculate these values, the interface elements must be assigned to the substructures and the interface equations must to be renumbered. Performing these operations for every partitioning obtained during the iterations would consume a significant amount of time. For that reason, the approximate values of the number of equations and the bandwidth are calculated by using the edge-cut (Section 2.2). The value of edge-cut shows the number of edges at substructure interfaces and is usually considered as an indication of the communication volume [1, 17, 29]. Both partitioning and repartitioning algorithms compute the edge-cut for each partitioning.

Figure 6.2 shows the edges at the interface of two substructures composed of 2D quadrilateral elements. As can be seen from the figure, there are three edges per interior node. Thus, in order to calculate the number of interface nodes of substructures composed of 2D elements, the edge-cut value is divided by three. Then, the number of interface equations is simply computed by multiplying the number of nodes with the number of degrees of freedom per node. When a structure is composed of 3D elements, there are six edges per node. Hence, for such structures the edge-cut value is divided by six to obtain the number of interface nodes.

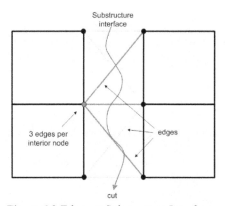

Figure 6.2 Edges at Substructure Interfaces

The other variable that is needed for the interface solution time estimation is the bandwidth. The actual bandwidth can only be known after the interface equations are renumbered. This process will take significant amount of time if performed for each partitioning. For that reason, an approximate bandwidth is computed using the following approach.

The bandwidth of the interface stiffness matrix mainly depends on the number of substructures. As the number of substructures increases, the bandwidth decreases. For that reason, the bandwidth of the interface stiffness matrix is first assumed to be constant and equal to a predetermined percentage of the number of interface equations whose value decreases as the number of processors increases. The following percentages that are shown in Table 6.1 were utilized to calculate the bandwidth by multiplying them with the number of interface equations. These values were based on the properties of the interface stiffness matrices of the example test problems.

Table 6.1 Bandwidth-Number of Interface Equation Ratios with Number of Processors

# Procs	Percentage	# Procs	Percentage
2	%100	4	%85
6	%70	8	%60
12	%50		

Once the number of interface equations and the bandwidth of a partitioning is approximately computed, the interface solution time is estimated by adding the estimated computation and communication costs of the parallel variable band solver. The details of the interface solution time estimation are explained in Chapter 5. Then, the condensation time is estimated by multiplying the governing operation count for condensation with the speed of inner-product. Finally, the condensation and the interface solution time estimations are added and the estimate for the total solution time is obtained.

The total time estimations are computed for each partitioning and the fastest one is chosen for solution. Then, the interface elements are assigned and the substructures are created. As a final step, the interface equations are numbered by using the bandwidth minimization algorithm [14].

The final results are written to a file which will be used by the parallel solution algorithm. The output file involves the nodes, elements, loadings of each substructure, material properties, boundary conditions and the interface information. The nodes of each substructure are written according to their optimized order so that the parallel solution program does not perform equation renumbering for condensation. The interface information describes the adjacency relationship of substructures. The final information in the output file is the numbering of the interface nodes.

6.2.2 Parallel Solution Algorithm, Multiple Loading Conditions

The algorithm for the parallel solution of problems with multiple loading conditions is given in Figure 6.3. The solution starts by creating separate databases at each processor. The master processor reads the input file prepared by the first program and sends the nodal, the element connectivity, and the loading informations of each substructure to the corresponding processor. The number of substructures is equal to the number of processors. Then, each processor creates substructure databases in their local memory. The details of this database are explained in detail at Appendix B.

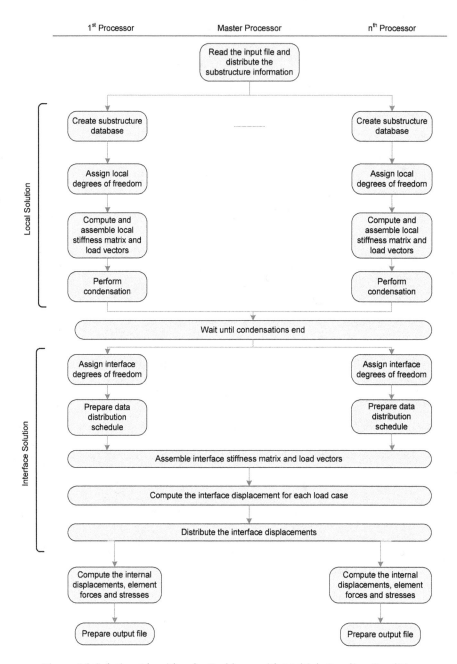

Figure 6.3 Solution Algorithm for Problems with Multiple Loading Conditions

After the creation of databases, the local solution starts. During the local solution, there is no need for communication between the processors. Moreover, the processors do not have to wait for each other in order to proceed to the next step of the local solution. First, each processor assigns the local degrees of freedom to the nodes of their substructures simultaneously. The nodes of each substructure were written into the input file according to their optimized order by the first program. The nodes are placed into the database in the same order. Hence, during the assignment process, each node is visited one by one and their active degrees of freedom are numbered consecutively. The nodes can have different number of active degrees of freedom.

The next step is the computation and assembly of the substructure level stiffness matrix and force vectors. Each processor computes and assembles the stiffness matrices of their elements. The force vectors are constructed in the same manner. Whenever the processors finish assembly, they start condensing both the stiffness matrix and the force vectors. No communication is required during assembly and condensation.

If the size of the stiffness matrix exceeds the size of the in-core memory, the out-of-core version of the assembly and the condensation algorithms will be utilized. The out-of-core assembly algorithm works in the following way: initially, the first 'n' columns of the stiffness matrix are assembled using the in-core memory. The 'n' is chosen in such a way that the size of the partial stiffness matrix does not exceed the available in-core memory. Next, the assembled part is written to the file and the assembly of the columns from 'n+1' to '2n' is initiated. This process continues until the whole matrix is assembled. Then, the matrix is condensed with the out-of-core version of the condensation algorithm explained in Section 3.6.

217

The interface solution does not start until all processors finalize the condensations since their results will be utilized during the interface solution. During the interface solution, the processors highly depend on each other. In other words, they not only send data to each other, but also wait for each other in order to proceed to the next step of the computation. Once all processors finish condensation, the interface degrees of freedom are assigned to the interface nodes. Likewise, the interface nodes can have different number of active degrees of freedom. Their order is also optimized by the first program and written to the input file. Thus, the numbering is performed according to that order to minimize the bandwidth.

The interface equations are solved in parallel and the interface equations must be distributed among processors. Currently, the rows of the stiffness matrix are assigned to the processors in a cyclic manner. In other words, if a processor keeps the first row, it will also keep the '$1+p^{th}$', '$1+2p^{th}$'... rows of the stiffness matrix where 'p' is equal to the total number of processors.

The assembly process is performed row by row in parallel. The rows are assembled one-by-one where each processor sends their contribution to the ith row of the stiffness matrix and force vectors to the processor that is assigned to store the ith row.

The interface displacements are computed by using the parallel variable band-solver as described in Chapter 5. At the end of the solution, each processor will have the displacements of their assigned rows. However, they need to have the displacements of their interface nodes in order to recover the internal displacements. Thus, at the end of the solution, the interface displacements are distributed to the corresponding processors.

The last step of the parallel solution is the recovery phase. Each processor computes their internal displacements for each loading condition simultaneously. Finally, the element forces and stresses are computed and the results are written to a file for post-processing.

6.2.3 Numbering

The substructure based parallel solution algorithm requires various numbering and data storage schemes for the different steps of the solution. The interface equations have different numberings for the local solution, interface solution and the way in which they are stored in each processor's local memory.

During the local solution, each processor numbers degrees of freedom of its substructures in such a way that the profile is minimized and the interface equations are stored at the end of the stiffness matrix. This means that each interface node will have a different equation numbering at each substructure. An example of local numbering of two substructures is presented in Figure 6.4:

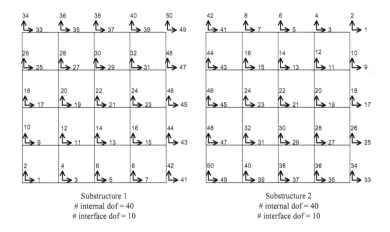

Figure 6.4 Local Numbering of Substructures

219

After condensation, the interface stiffness matrix must be assembled. For that purpose, the interface degrees of freedom are assigned to the interface nodes. In this process, the substructure interfaces are joined and the same degrees of freedom are assigned to the interface nodes which have the same coordinates. An example of the interface numbering is shown in Figure 6.5:

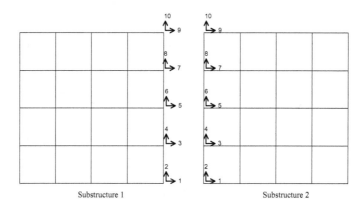

Figure 6.5 Interface Numbering of Substructures

Since the interface equations are solved by using the parallel variable band solver, the interface stiffness matrix must be distributed among processors. When the interface stiffness matrix is distributed in row-wise manner, each processor stores different rows of the interface stiffness matrix at their local memory. This means there is a need for another numbering scheme that maps the rows of the interface stiffness matrix to the rows of the distributed matrices as shown in Figure 6.6:

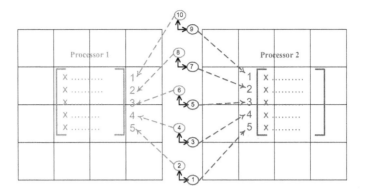

Figure 6.6 Distributed Numbering of Interface Stiffness Matrix

In order to simplify the handling of various numbering schemes discussed above, two list classes were created. The first one, called DistributionList Class, describes how the interface stiffness matrix is distributed among processors. The output of the Data Distribution List object for the above structure is given in Table 6.2.

Table 6.2 Data Distribution List

Interface #	# Data Holders	Data Holders	Owner
1	2	1,2	2
2	2	1,2	1
3	2	1,2	2
4	2	1,2	1
5	2	1,2	2
6	2	1,2	1
7	2	1,2	2
8	2	1,2	1
9	2	1,2	2
10	2	1,2	1

The Data Distribution List is composed of four columns. The first column shows the interface equation numbers. The second and the third columns show the number and the ids of processors which have a contribution to that interface equation, respectively. The last column shows the id of the owner processor that will store that interface equation. In

221

this list, it is assumed that the substructures and their assigned processors have the same ids.

The second list class, called DofMapper, stores all the ids of an interface equation assigned by different numbering schemes and provides conversion routines between these numbering schemes. The stored numberings for the above example problem is given in Table 6.3. The first column shows the local numberings of the interface equations that is used by the condensation algorithm. The second column shows the row numbers of each interface degrees of freedom whereas the third column shows the row numbers of the interface equations when they are distributed among the processors.

Table 6.3 Dof Mapper Lists for Each Processor

Processor 1				Processor 2		
Local	Interface	Dist.		Local	Interface	Dist.
41	1	0		41	10	0
42	2	5		42	9	1
43	3	0		43	8	0
44	4	4		44	7	2
45	5	0		45	6	0
46	6	3		46	5	3
47	7	0		47	4	0
48	8	2		48	3	4
49	9	0		49	2	0
50	10	1		50	1	5

Prior to the interface solution, the DofMapper and DistributionList classes are prepared. First, the local degrees of freedom are assigned to the interface nodes. After the condensation, the interface degrees of freedom are assigned. At the same time, the first and the second columns of each substructure's DofMapper objects are filled. This way, the mapping between the local degrees of freedom and interface degrees of freedom is provided. At the same time, the first and the second columns of the DistributionList object is also filled. After that point, the distribution schedule is prepared by simply filling the third column of the DistributionList object in cyclic manner. As a final step, the third column of DofMapper object that provides mapping information between the interface degrees of freedom and their distributed locations is filled. As a result, the DofMapper objects of each processor have different values for each processor whereas the DistributionList objects are the same at all processors. Both objects are utilized by the interface stiffness matrix assembly, interface solution, and interface displacement distribution algorithms for mapping and data distribution.

6.2.4 Parallel Assembly

The rows of the interface stiffness matrix must be assembled and distributed to the processors before the solution. The stiffness matrix is assembled row-by-row in parallel starting from the first row. The pseudo code of the parallel assembly algorithm is shown in Figure 6.7:

```
for i = 1 to number of interface equations {
  if (owner of i) {
    for j = 1 to number of data holders {
      Receive column Ids from jth data holder
      Receive contributions to the ith row from jth data holder
      Assemble
    }
  } else if (data holder for i) {
    Prepare data to send
    Send column Ids to the owner of i
    Send contributions to the ith row to the owner of i
  }

  Wait All Other Processors
}
```

Figure 6.7 Pseudo Code for Parallel Assembly Algorithm

223

The assembly starts from the first row. The processor that will store the first row of the stiffness matrix, called the owner of the first row, will receive data from other processors which have a contribution to the first row. The processors that do not have any data for the first row wait until all send and receive operations for the first row are completed.

The processors that will send data to the owner of the first row prepare two separate data buffers. The first buffer contains the column ids that will be used to assemble the stiffness values to the first row. The second buffer contains the stiffness values. Then, both buffers are sent to the processor that will store the first row.

The processor which stores the first row can receive data from only a single processor at a time. Thus, the other processors are queued and wait their turn to send their data. Every time the owner processor finishes the receive operation, it assembles the incoming data and uses the same buffer for the new incoming data. This procedure is repeated for all rows of the interface stiffness matrix.

The DistributionList objects facilitate the above operations significantly. The processor that will store a current row of the stiffness matrix is determined by the DistributionList. The processor that will send data obtains the id of the owner processor from the DistributionList. Similarly, the owner processor obtains the ids of the processors that will send data from the DistributionList. Moreover, if any other data distribution scheme other than a cyclic scheme is utilized, only the third column of the DistributionList will be modified.

6.3 Results and Discussions

The test runs were performed on the clusters in one of the student computer labs of CEE at Georgia Tech. The lab had 12 identical computers each having 2.4 GHz Pentium 4 processors with 512 Kbytes cache memory and 1 Gbyte memory with a bus speed of 800 MHz. All computers were running Windows XP Professional. The computers were connected to the school's network with 1 GBit network cards.

The efficiency of this framework was tested on several structural models obtained from AE firms (Appendix A). Two graphs are presented for each model. For each model, the graph at the top shows the total solution time, from stiffness matrix generation to internal displacement computation. The graph at the bottom shows the speed-up values that were calculated by dividing an estimation of the in-core serial solution time with the parallel solution times. The serial time estimation considered only the factorization phase and was determined by multiplying the operation count with the speed for inner product operation. This way, the effect of any inefficiency due to out-of-core solution during the serial solution was eliminated.

6.3.1 2D Square Mesh

The first example is the Square 2D Mesh (Appendix A.1.1) modeled with shell elements. The model was composed of 25,600 quadrilateral shell elements with 26,266 nodes and 157,578 equations.

First, the effect of workload balancing step on the interface solution time was examined. The interface stiffness matrices of the initial substructures obtained using METIS [30], and the substructures balanced with both the diffusion and scratch-remap algorithm from PARMETIS [28] were solved with the parallel variable band solver as described in Chapter 5. Eight processors were utilized for the solution.

The following three Figures 6.8 to 6.10 show the partitioning, interface stiffness matrix fill-in and the interface solution times of the initial substructures, substructures created with diffusion and scratch-remap algorithms, respectively. The black areas in the stiffness matrices represent the non-zero terms before factorization.

As can be seen from Figure 6.8, the initial substructures resulted in the smallest interface problem. Not only did the initial substructures have the smallest number of interface equations, but also they also had the smallest bandwidth. The solution time of substructures created with the scratch-remap algorithm was only 4 seconds slower than the initial substructures. On the other hand, diffusion algorithm created a larger interface problem which required 33 seconds for the solution. The substructures created with both diffusion and scratch-remap algorithms had larger bandwidths than the initial structures because the substructures became more interconnected to each other after the iterations. For example, although the second substructure had three neighbor substructures at the initial partitioning, this number increased to 5 after workload balancing with scratch-remap algorithm. That caused an increase in the bandwidth of the interface stiffness matrix.

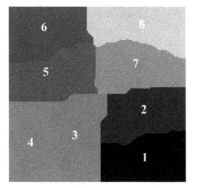

Substructures   Stiffness Matrix Fill-in

Number of Equations	Average Bandwidth	Solution Time
4398	2094	20.32 secs

Figure 6.8 Interface Solution with Initial Substructures

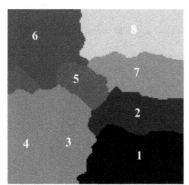

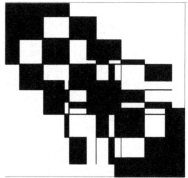

Substructures   Stiffness Matrix Fill-in

Number of Equations	Average Bandwidth	Solution Time
5478	3144	32.98 secs

Figure 6.9 Interface Solution with Substructures Created with Diffusion Algorithm

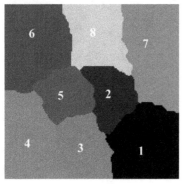

Substructures

Stiffness Matrix Fill-in

Number of Equations	Average Bandwidth	Solution Time
4632	2502	23.97 secs

Figure 6.10 Interface Solution with Substructures Formed with Scratch-remap Algorithm

The model was then solved with a single loading condition using 2 to 12 processors with each partitioning, the initial substructures and the substructures balanced with diffusion and scratch-remap type repartitioning algorithms. For all cases, the local equations were renumbered by using the bandwidth minimization algorithm [14].

The parallel solution times and the speed-up values for each partitioning are presented in Figure 6.11. When the same problem was solved with GTSTRUDL [122], a structural analysis and design software, 763 seconds was spent for the solution and 180 seconds was spent for stiffness matrix generation. The serial solution time for this problem was estimated as 869 seconds.

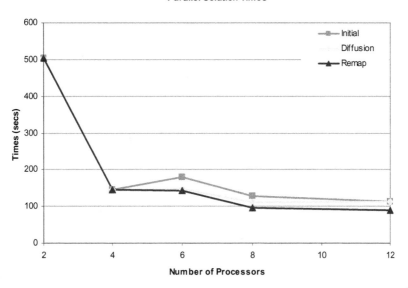

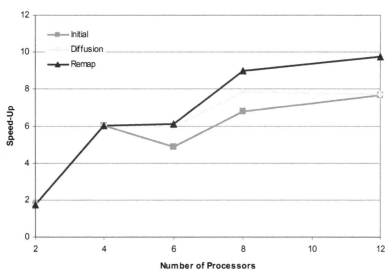

Figure 6.11 Parallel Solution Times and Speed-up Values for 2D Square Mesh Problem

The condensation times of the substructures were balanced when the structure was partitioned into 2 and 4 substructures. When the structure was partitioned into more than 4 substructures, a significant imbalance occurred in the condensation times of the initial substructures. When the condensation times were balanced, the total solution time was decreased.

As can be seen from Figure 6.11, the parallel solution with the substructures balanced with the scratch-remap algorithm performed much faster (15% with 8 processors, 26% with 12 processors) than the substructures balanced with the diffusion algorithm. This is mainly because the scratch-remap algorithm resulted in a smaller interface problem than the diffusion algorithm. Moreover, the speed-up increased as the number of processors increased for substructures balanced with scratch-remap algorithm. The solution time reduced to 89.1 seconds when 12 processors were utilized.

Table 6.4 shows the condensation, interface equation assembly, and interface solution times of each partitioning when 12 processors were utilized. The governing condensation time of the initial substructures were 62.95 seconds whereas the governing condensation time reduced to 47.16 and 36.44 seconds for the substructures balanced with diffusion and scratch-remap algorithms, respectively. On the other hand, the initial substructures had the fastest interface assembly and the solution times. The diffusion algorithm increased the interface solution time significantly and more than the effect of the gain in the condensation time resulting in a slower the total solution time. The substructures balanced with the scratch-remap algorithm increased the interface solution time by only 2.4 seconds but decreased the condensation time by 26.5 seconds.

Table 6.4 Solution with 12-processors, Single Loading Condition (seconds)

	Condensation	Parallel Assembly	Interface Solution	Sum	% Change from Initial
Initial	62.95	8.58	40.57	112.10	-
Diffusion	47.16	10.56	53.90	111.62	-0.4%
Scratch-Remap	36.44	8.80	42.97	88.21	-21.3%

Table 6.5 and Table 6.6 show the itemized parallel solution times with 100 loading conditions for the initial substructures and the substructures balanced with scratch-remap algorithm, respectively. The values under the 'Local Solution' column show the timing of the local solution which is composed of element stiffness generation and assembly, incomplete stiffness matrix factorization, and forward and back substitution. The values under the 'Interface Solution' column show the timing for the interface solution which is composed of assembly of interface equations in parallel, factorization of interface stiffness matrix, forward and back substitution of the interface load vectors, and distribution of the interface displacements.

Table 6.5 Solution with Initial Substructures, 100 Loading Conditions (seconds)

# SS	Local Solution				Interface Solution				Total
	A	F	FS	BS	PA	F+FS	BS	DD	
2	9.17	498.58	92.29	37.96	1.04	1.34	0.16	0.39	640.93
4	4.67	136.19	33.96	10.62	1.56	5.19	0.30	0.41	192.90
6	3.05	158.83	28.49	8.20	3.11	16.68	2.55	0.47	221.38
8	2.25	101.64	19.41	5.63	4.78	19.26	4.63	0.49	158.09

A: assembly, PA: parallel assembly, F: Factorization, FS: Forward substitution, BS=Back substitution, DD: Data distribution

As can be seen from Table 6.5, the local factorization time with 6 substructures were more than the local factorization time with 4 substructures. This is mainly due to the

231

increased number of interface nodes with much higher column heights when the structure

was partitioned into six substructures. The time spent during assembly and the forward

and back substitution steps of the local solution decreased as the number of processors

increased. On the other hand, the interface solution time increased as the number of

processors increased due to an increase in the size of the interface as the number of

substructures increased. The parallel assembly and back substitution steps consumed 33%

of the interface solution time.

Table 6.6 Solution with Balanced Substructures, 100 Loading Conditions (seconds)

# SS	Local Solution				Interface Solution				Total
	A	F	FS	BS	PA	F+FS	BS	DD	
2	9.17	498.58	92.29	37.96	1.04	1.34	0.16	0.39	640.93
4	4.67	136.19	33.96	10.62	1.56	5.19	0.30	0.41	192.90
6	3.83	113.27	27.79	8.21	4.56	25.83	3.45	0.60	187.54
8	2.94	65.19	18.85	5.56	5.08	25.71	3.24	0.53	127.10

A: assembly, PA: parallel assembly, F: Factorization, FS: Forward substitution, BS=Back substitution, DD: Data distribution

When the results in Table 6.6 are compared with the results in Table 6.5, the local

factorization times of the balanced substructures were 45.56 and 36.45 seconds faster

than the local factorization times of the initial substructures for the solution with 6 and 8

processors, respectively. However, workload balancing step resulted in an increase in the

interface solution by 11.5 and 5.4 seconds for the 6 and 8 processor solutions,

respectively. As a result, balancing the condensation times of the initial substructures

decreased the total solution time.

6.3.2 Half-Disk

The second model is the Half Disk Model (Appendix A.1.2) which is composed of 28,128 brick elements and 36,773 nodes. The parallel solutions were performed using 2, 4, 6, 8, and 12 computers and the performances of the initial substructures and the substructures balanced using both the diffusion and scratch-remap algorithms were compared. The local equations were renumbered using the profile reduction algorithm [15].

Figure 6.12 shows the parallel solution results with a single loading condition. The estimated serial solution time for this model was computed as 1140 seconds. GTSTRUDL [122] required 1084 seconds for solution and 54 seconds for element stiffness generation and assembly.

The condensation times of the initial substructures were not balanced for all cases, even for the solution with two processors. Thus, the workload balancing step decreased the total solution time for all cases. The substructures balanced with diffusion algorithm had the fastest solution times when 2 and 4 processors were utilized for the solution. Their efficiency decreased as the number of processors increased. The substructures balanced with scratch-remap algorithm performed the fastest when more than 4 processors were utilized. The total solution time was 34%, 47%, and 26% faster when compared with the results of substructures balanced with diffusion algorithm using 6, 8, and 12 processors respectively. This is mainly due to having larger interface problem when diffusion algorithm was used.

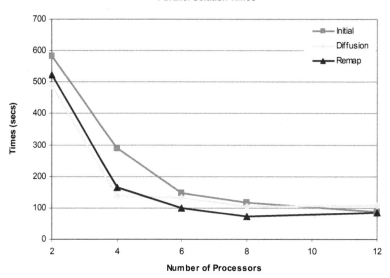

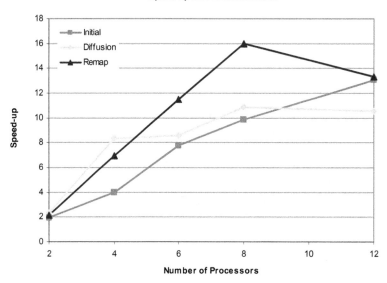

Figure 6.12 Parallel Solution Times and Speed-up Values for Half Disk Problem

234

The fastest solution time was obtained with 8 processors using the substructures balanced with the scratch-remap algorithm. On the other hand, the solution time decreased when 12 processors was utilized. This is mainly because the size of the substructures was reduced so much that the interface solution started to govern the total solution time. Since the interface solution algorithm was not very scalable for the DELL cluster due to its slow communication speed as discussed in Section 5.3.6, the total solution time decreased when more processors were utilized.

The detailed solution times of the initial substructures and the substructures balanced with both the diffusion and scratch-remap algorithms using 12 processors are shown in Table 6.7. The diffusion based balancing did not decrease the condensation time much (7%). In addition to that, diffusion based balancing caused a significant increase in the interface solution time (80%). That's why the total solution time of substructures balanced with diffusion algorithm was slower than the initial substructures. On the other hand, the scratch-remap based balancing did not increase the interface solution time significantly (10%) and thus was able to decrease the total solution time.

Table 6.7 Solution with 12-processors, Single Loading Condition (seconds)

	Condensation	Parallel Assembly	Interface Solution	Sum	% Change from Initial
Initial	53.98	7.49	27.51	88.98	-
Diffusion	50.22	9.50	49.70	109.42	+23%
Scratch-Remap	40.41	8.23	30.38	79.02	-11%

The same model was solved again for 100 multiple loading conditions by using the initial substructures and the substructures balanced with the scratch-remap algorithm. Tables 6.8 and 6.9 present the itemized timing for both cases.

Table 6.8 Solution with Initial Substructures, 100 Loading Conditions (seconds)

# SS	Local Solution				Interface Solution				Total
	A	F	FS	BS	PA	F+FS	BS	DD	
2	21.09	576.88	85.23	28.43	0.81	1.00	0.42	0.34	714.20
4	10.17	283.76	42.46	11.98	1.49	4.86	1.65	0.40	356.77
6	6.89	135.75	23.37	6.65	2.46	10.96	2.08	0.40	188.56
8	5.15	104.62	17.16	4.89	2.82	12.10	3.35	0.34	150.43

A: assembly, PA: parallel assembly, F: Factorization, FS: Forward substitution, BS=Back substitution, DD: Displacement distribution

Table 6.9 Solution with Balanced Substructures, 100 Loading Conditions (seconds)

# SS	Local Solution				Interface Solution				Total
	A	F	FS	BS	PA	F+FS	BS	DD	
2	21.88	521.93	85.36	24.03	0.66	0.78	0.35	0.30	655.39
4	13.24	157.35	36.53	10.45	1.40	3.66	1.40	0.33	224.36
6	9.00	85.61	19.89	5.72	2.78	11.47	2.32	0.42	137.21
8	6.35	53.39	14.05	4.10	3.76	12.28	3.55	0.38	97.86

A: assembly, PA: parallel assembly, F: Factorization, FS: Forward substitution, BS=Back substitution, DD: Displacement distribution

As can be seen from Tables 6.8 to 6.9, the solution with the balanced substructures had the smaller condensation times. The workload balancing not only decreased the local factorization time, but also the local forward and back substitution times. On the other hand, the local stiffness matrix assembly time increased due to unequal number of elements in each substructure. Although there were 100 loading conditions, the time required for interface displacement distribution was negligible. Moreover, the interface problem size did not increase much with the balanced substructures. As a result, the workload balancing with the scratch-remap algorithm decreased the solution times significantly (e.g. %59 with 4 processors, %54 with 8 processors).

6.3.3 Bridge Deck

The third model is the Bridge Deck model (Appendix A.1.3) composed of 24,200 brick elements and 34,239 nodes. The parallel solutions were performed using 2, 4, 6, 8, and 12 computers and the performances of the initial substructures and the substructures balanced with the diffusion algorithm were compared. The equations were numbered using the profile reduction algorithm.

The results of the parallel solutions with a single loading condition are presented in Figure 6.13. The estimated serial solution time was 659 seconds. GTSTRUDL [122] performed the solution in 665 seconds and element generation and assembly in 246 seconds.

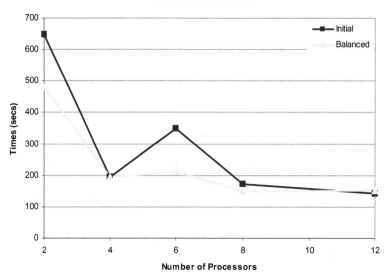

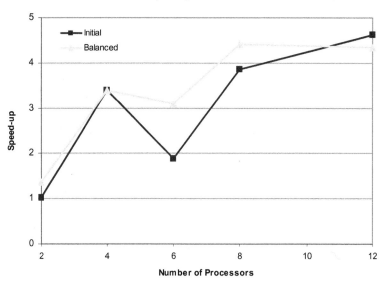

Figure 6.13 Parallel Solution Times and Speed-up Values for Bridge Deck Problem

238

In this problem, the parallel solution with balanced substructures was faster than parallel solution with the initial substructures up to 8 processors. The 12 processor solution with balanced substructures was slower than the parallel solution with initial substructures because workload balancing increased the interface solution time more than it decreased the condensation time.

The speed-up values obtained for the parallel solution of the bridge deck model was smaller when compared with the previous examples. For example, the 4 and 8 processor solution of the 2D square mesh model decreased the serial solution time by 6 and 9 times, respectively. Similarly, for the disk structure, the 4 and 8 processor solutions decreased the serial solution time by 6 and 14 times, respectively. This is mainly because the substructuring not only decreased the size but also the bandwidth of the local problem. Although the interface problem size also increases as the number of substructures increases, the solution time was mostly governed by the local solution.

On the other hand, in the bridge deck model, the maximum speed-up was obtained with the initial substructures using 12 processors which was equal to 4.5. This is a rather low value when compared with the previous examples. In this specific example, the partitioning decreased the number of equations in each substructure but not the profile of their stiffness matrix. Thus, the speed-up in the local solution time was lower than the number of processors because of the increased column heights at the interface equations. When the cost of the interface solution was added to the low local solution time improvement, rather small speed-up values were obtained for the total solution time.

6.3.4 High-Rise Building I

Up to this point, all the previous examples were composed of one type of elements, either shell or brick elements. In these example problems, the workload balancing step was able to decrease the local solution up to 50%. Moreover, the solution could be performed using the in-core memory only.

The High-rise Building I model, on the other hand, is composed of 1,221 frame and 61,307 shell elements with 54,752 nodes. The parallel solutions were performed by using 2, 4, 6, 8, and 12 computers and the performances of the initial substructures and the substructures obtained after the workload balancing with scratch-remap algorithm were compared. The local equations were renumbered using the profile reduction algorithm [15]. When the serial solution is performed, the size of the stiffness matrix was more than 8 Gbytes which exceeded the available in-core memory. As a result, the out-of-core local solver was utilized for the condensation using 2, 4 and 6 processors.

Figure 6.14 shows the parallel solution times of high-rise building model for 100 loading conditions. The estimated serial solution time was computed as 40,305 seconds. As can be seen from the figure, the parallel solution with balanced substructures was faster. The reduction in the solution time was 1366, 620 and 383 seconds for 4, 6, and 8 processor solutions, respectively. The workload balancing step did not provide an improvement in the total solution time when 12 processors were utilized due to a larger interface problem.

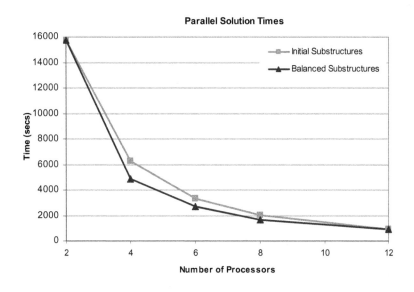

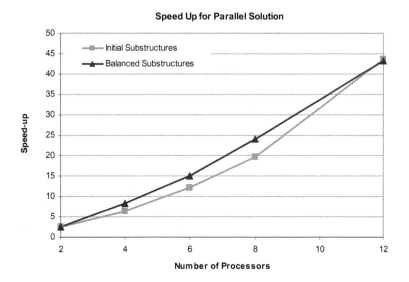

Figure 6.14 Parallel Solution Times and Speed-up Values for High-Rise Building I Problem

The parallel solution of this problem resulted in very-high speed up values. For example, the total solution time with 12 processors was 932 seconds whereas the serial solution time was 40,305 seconds. In other words, the parallel solution was 43 times faster than the serial solution. This indicates that partitioning also decreased the profile of the stiffness matrices of substructures.

Table 6.10 presents the itemized timings for the parallel solution. The 2, 4, and 6 processor solutions were performed by the out-of-core local solution algorithm. The out-of-core solver performs the factorization and the forward substitution together, thus the value under the local factorization column of Table 6.10 shows the total time spent during factorization and forward substitution up to 6 processor solution.

Table 6.10 Solution with Final Substructures, 100 Loading Conditions (seconds)

# SS	Local Solution				Interface Solution				Total
	A	F	FS	BS	PA	F+FS	BS	DD	
2	206.93	15069.18	-	424.70	24.11	4.72	1.09	0.66	15731.4
4	58.86	4570.43	-	137.74	66.84	48.45	5.85	1.02	4889.2
6	34.08	2397.47	-	76.78	82.88	86.14	8.00	1.06	2686.4
8	5.35	1369.31	121.69	45.25	21.25	93.67	12.00	1.04	1669.6
12	4.07	586.64	57.26	16.38	30.13	211.21	21.41	4.75	931.8

A: assembly, PA: parallel assembly, F: Factorization, FS: Forward substitution, BS=Back substitution, DD: Data distribution

As can be seen from Table 6.10, using out-of-core storage increased the assembly time. Although the interface problem size was not high for the 2 processor solution, the parallel assembly time was 5 times slower than the interface factorization time. This is mainly because of the cost of file I/O while reading the interface columns of the factorized substructure level stiffness matrix.

242

The same model was solved with varying number of loading conditions using 12 processors. The parallel solution was performed for 1, 50, 100, 250, 500, and 1000 loading conditions and the total solution times for each case are shown in Figure 6.15.

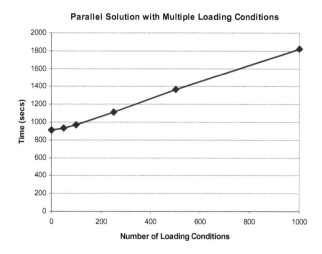

Figure 6.15 Parallel Solution Times with Multiple Loading Conditions for High-Rise Building I Model

As can be seen from Figure 6.15, there was a linear relationship between the solution time and the number of loading conditions. The solution time increased as the number of loading conditions was increased. When there was a single loading condition, the solution was finalized in 909 seconds; however, as the number of loading conditions was raised to 1000, the time required for the solution was almost two times larger, 1820 seconds. In other words, approximately, 0.9 seconds was spent to solve each additional loading condition.

Table 6.11 shows the detailed load factorization times for the parallel solutions. The local assembly and factorization times remained constant and equal to 4 and 588 seconds,

243

respectively, for all solutions. The local forward and back substitution times increased as the number of loading conditions was increased and the time spent during local forward substitutions was three times slower than the time spent during back substitutions.

In Table 6.11, the parallel assembly time was presented in two parts; parallel stiffness matrix and load vector assembly. The parallel stiffness assembly time remained almost constant for each case but the parallel load factorization time increased as the number of loading conditions was increased. Similarly, the cost of redistributing the interface displacements was also increased. When there were 1000 loading conditions, the parallel load vector assembly and redistribution of interface displacements consumed 7% of the total interface solution time.

When the number of loading conditions was varied from 100 to 1000, the time spent during factorization and forward substitution of interface equations increased. On the other hand, the time spent during factorization and forward substitution with a single loading condition was slower than the one with 100 loading cases. This was due to the variable broadcasting speed of the DELL cluster. When the solution was performed with a single loading condition, the communication cost of parallel factorization was larger than the communication cost with 100 loading conditions although the same amount of data was transferred.

The interface back substitution required 10.7 seconds when there was a single loading condition. Approximately 99% of this time was spent for data initiation for communication (start-up latency). As the number of loading conditions increased, the ratio of the start-up latency cost decreased and performing the back substitutions in parallel became more efficient. When the start-up latency cost was excluded, approximately 0.1 seconds was spent for back substitution of each additional loading condition.

Table 6.11 Load Factorization Times for Solution with Multiple Loading Conditions (seconds)

# Load	Local Load Factorization		Interface Solution					Total
	FS	BS	PSA	PLA	F+FS	BS	DD	
1	0.6	0.2	35.6	0.7	265.4	10.7	2.4	908.8
50	28.3	8.3	35.5	1.2	251.5	15.2	3.1	934.5
100	56.7	16.5	35.4	1.8	242.3	20.6	3.6	968.3
250	141.4	41.1	40.7	3.6	251.0	33.8	4.2	1110.0
500	282.9	84.6	35.2	7.6	273.7	74.9	15.8	1367.3
1000	567.2	167.4	41.0	14.2	308.5	108.1	21.6	1820.8

PSA: Stiffness assembly, PLA: Load assembly, F: Factorization, FS: Forward substitution, BS=Back substitution, DD: Data distribution

6.3.5 High-Rise Building II

The High-rise Building II model is another actual structural model that has large geometrical irregularities in the lower stories. It is composed of 6,001 frame members and 54,450 quadrilateral and triangular shell elements with 54,773 nodes. The model was solved by using both the initial substructures and the substructures balanced with scratch-remap algorithm. The profile reduction algorithm [15] was utilized for the local equation renumbering. The solutions with 2, 4, and 6 processors utilized the out-of-core version of the local solution algorithm.

Figure 6.16 shows the parallel solution with 100 loading conditions. The solutions with the balanced substructures were 863, 454, and 165 seconds faster than the solution with the initial substructures for 4, 6, and 12 processor solutions, respectively. The maximum speed-up was obtained with balanced substructures and equal to 41.

245

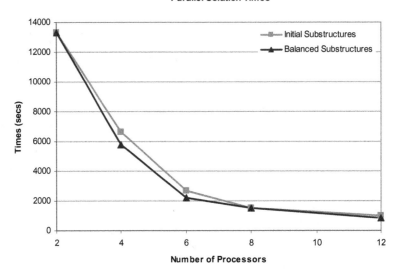

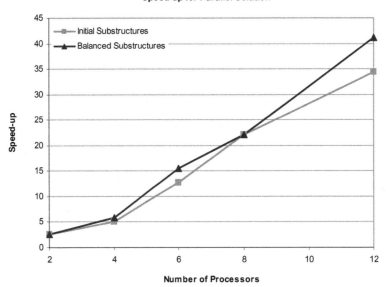

Figure 6.16 Parallel Solution Times and Speed-up Values for High-Rise Building II Problem

246

6.3.6 Nuclear Waste Plant

The last example model is the Nuclear Waste Plant model. It is composed of 2,811 frame members and 43,776 quadrilateral shell elements with 39,440 nodes. The problem was also solved using the initial substructures and the substructures balanced with the scratch-remap algorithm. The profile reduction algorithm [15] was utilized for the local equation numbering. The solutions with 2, 4, and 6 processors utilized the out-of-core version of the local solution algorithm.

Figure 6.17 shows the parallel solution times with 100 loading conditions. In this problem, the balanced substructures did not decrease the solution time as much as the previous examples (at most 10%). This is mainly because of the small improvement in the local solution times and having a larger interface problem after the workload balancing step. Still, the gain in the total solution time was 607 and 104 seconds for 4 and 8 processor solutions, respectively. No additional improvement was obtained for the 6 and 12 processor solutions.

The estimated solution time was computed as 32,564 seconds. The parallel solution with 12 processors was able to decrease the serial solution time to 1,393 seconds. In other words, the maximum speed-up with 12 processors was 24.5.

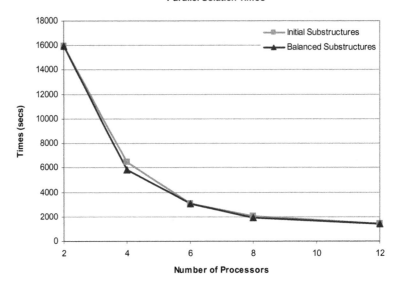

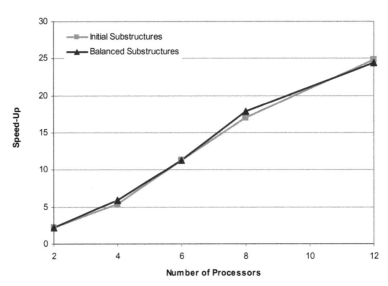

Figure 6.17 Parallel Solution Times and Speed-up Values for Nuclear Waste Plant
Problem

248

6.4 Conclusions

This study presented a substructure based parallel linear solution framework for the static analysis of linear structural engineering problems having multiple loading conditions. The framework was composed of two separate programs designed to work on PC Clusters having the Windows operating system. The first program was responsible for creating the optimum substructures for the parallel solution. First, the structures were partitioned in such a way that the number of substructures was equal to the number of processors. Then, the estimated condensation time imbalance of the initial substructures was adjusted by iteratively transferring nodes from the substructures with slower estimated condensation times to the substructures with faster estimated condensation times. In order to decide which nodes were needed to be transferred, either the diffusion or scratch-remap type repartitioning algorithms was utilized. Once the final substructures were obtained, the second program initiated the solution. Each processor assembled its substructure's stiffness matrix and condensed it to the interface with other substructures. The interface problem was solved by the parallel variable band solver. After computing the interface unknowns, each processor calculated the internal displacements. Examples which illustrated the applicability and efficiency of this approach were presented. In these examples, the number of processors was varied from one to twelve to demonstrate the performance of the overall solution framework.

Balancing the condensation times of substructures decreased the total parallel solution time for most of the example problems. The workload balancing was able to decrease the not only the local factorization time but also the local forward and back substitution times. On the other hand, there were some cases where although the local solution times was decreased by the workload balancing step, the interface problem size was increased so much that there was not any gain in the total solution time. This situation most frequently occurred for solutions with 12 processors.

The substructures balanced with the scratch-remap algorithm performed better than the substructures balanced with the diffusion algorithm. The scratch-remap algorithm not only produced the fastest local solution time but also it created smaller interface problems especially when the number of processors was more than 4. The initial substructures had the smallest interface problem size for almost every case but the difference between the interface solution time of the initial substructures and the substructures balanced with the scratch-remap algorithm was not high. Since the total solution time was governed by the local solution time, the substructures balanced with the scratch-remap algorithm performed the fastest.

The total solution time decreased as the number of processors increased. The solution produced larger speed-up as the size of the problem increased. This is mainly because the solution time of the large problems tested in this study was primarily governed by the local solution time. Thus, substructuring not only decreased the number of equations but also the profile of the stiffness matrix of the local problem. As a result, the speed-up were larger than the number of processors.

The presented method is very efficient for problems having multiple loading cases. It allows performing factorization of each load case in parallel with little increase in the communication overhead. Another advantage of this framework is the use of out-of-core solvers for the local solution when necessary. Thus, problems which exceed the in-core memory can be solved more efficiently. Overall, this framework is very suitable and can be utilized to solve large linear static problems with multiple loading conditions in parallel.

CHAPTER 7

SUMMARY AND RECOMMENDATIONS FOR

FUTURE WORK

7.1 Summary

This study presented a substructure based parallel solution framework for large linear systems having multiple loading conditions. Every step of the solution, from partitioning of the structures into substructures to the computation of the internal displacements in the substructures was performed in parallel.

The first chapter presented an overview of existing parallel computing architectures and a literature survey of existing parallel solution methods. Among the current parallel architectures, the PC-cluster systems were chosen to be the target parallel environment due to their availability in civil engineering design offices and their cost despite their relatively low communication speed between processors when compared with other parallel architectures. Similarly, among many existing parallel solution approaches, a substructure based solution method was chosen to be the most suitable method for this study since such methods not only decreased the communication cost but also allowed performing the stiffness and force matrix generation, assembly, and computation of element results in parallel.

The first step of any substructuring method is to divide the structure into smaller substructures which is generally performed by partitioning algorithms. Thus, the second chapter focused on METIS [30] library which was utilized as the initial partitioning algorithm in this framework. The partitioning methods utilized in METIS [30] library were heuristic methods that were based on a multilevel approach which created quality partitions in a very short amount of time. The framework also utilized the PARMETIS [28] library, which had the parallel implementations of both the scratch-remap and diffusion type repartitioning algorithms, in order to repartition the structure to balance the condensation times of the substructures. Hence, a brief discussion regarding the repartitioning methods was presented.

Chapter 3 focused on the condensation algorithm. In this study, an active-column solver was utilized for condensation. Such solvers require the interface equations to be assembled last. For that reason, the internal equations were numbered by using a profile minimization algorithm and the interface equations were numbered after the non-interface equations. Due to this numbering, the column heights of the interface equations were much higher. The analytical examination of the condensation algorithm indicated that the number of interface equations significantly affected the condensation time. Moreover, the interface equations were one of the main sources of condensation time imbalance of substructures since none of the partitioning algorithms were able to balance the number of interface equations or their column heights.

The condensation time imbalance of substructures significantly decreases the efficiency of any substructure based parallel solver because the interface solution could not start until all condensations are completed. It was observed that the condensation times of substructures created with the existing partitioning algorithms differed considerably. Thus, Chapter 4 presented a workload balancing method developed to balance the condensation times of substructures. The workload balancing method was based on iteratively transferring nodes from substructures with slower estimated condensation times to the substructures with faster estimated condensation times. The repartitioning algorithms from the PARMETIS [28] library were utilized to select the nodes to transfer to adjacent substructures in order to decrease the imbalance. The method was tested on various example problems and the effects of the type of repartitioning and equation numbering algorithms were examined. At present, the workload balancing algorithm is limited to homogeneous PC clusters.

The workload balancing step decreased the condensation times in almost every case. Moreover, the time spent during the iterations was insignificant when compared with the improvement in the condensation times. Thus, one of the main contributions of this study was the development of a workload balancing method for direct condensation that was fast enough to be utilized prior to the actual linear static analysis, effective enough to decrease the total parallel solution time, and robust enough to work with mixed structural models (1D members mixed with 2D shell elements). Furthermore, it was also shown that the workload balancing with the scratch-remap based repartitioning algorithm created substructures which exhibited faster condensation times and smaller interface problem size when compared with the substructures created with the diffusion based repartitioning.

The solution was transferred to the substructure interfaces after the substructure level stiffness matrices had been condensed. The interface equations were assembled and solved in parallel to complete the solution. The interface solution was performed by a parallel variable band solver designed for problems with multiple loading conditions. The details of the solver were presented in Chapter 5. The solver utilized the row-wise formulation of LU decomposition method and performed both the factorization and forward substitutions in a manner which significantly decreased the communication cost. Analysis using three different PC clusters showed that the speed-up obtained with the parallel interface solution increased as the bandwidth of the interface stiffness matrix increased. Moreover, the scalability of the interface solution highly depended on the communication vs. computation speed ratios of the parallel environment. As this ratio increased, the interface solution became more scalable.

Chapter 6 presented the implementation of the substructure based parallel solution framework which consisted of two programs. The first program was responsible from data preparation for the parallel solution that included partitioning, workload balancing, and equation numbering. The second program was a parallel finite element program that performed the substructure based parallel solution. Both programs were written with C++ and FORTRAN programming languages and utilized MPICH [115] for parallelization.

Then, Chapter 6 focused on the framework's parallel efficiency after briefly describing both programs. Various structural models were solved on a PC cluster systems. The results demonstrated that balancing the condensation times of substructures also decreased the total solution time in almost every case. For most of the problems, the total solution time decreased as the number of processors increased. Moreover, substructuring not only decreased the number of equations but also the profile of the substructures' stiffness matrices resulting in speed-up values greater than the number of processors.

One of the main problems of solution frameworks using direct solvers is insufficient in-core memory. When the size of the stiffness matrix exceeds the available in-core memory, paging starts and the performance of the solution drops significantly. For such cases, either the number of processors is increased or the out-of core solvers are utilized. The structure of this framework allowed the utilization of the out-of-core solvers very efficiently in parallel. For example, the time spent for the out-of-core condensation was only 35% slower than the in-core condensation when solving the High-rise Building Model I using 4 processors. Thus, the other important contribution of this study is to present an efficient solution framework that can solve models of any size without requiring more processors than those which are available.

Normally, when a structure is analyzed serially, the cost of stiffness matrix assembly and the computation of element forces and stresses are much smaller than the cost of the solution. However, if a parallel solution algorithm focuses on decreasing the solution time only, the assembly and element result computations may govern the total solution time. For example, if the stiffness matrix of the Half-Disk model was assembled serially, the stiffness matrix generation and assembly would consume approximately 45 seconds which was 45% of the total solution time (92 seconds) when 8 processors were utilized. Hence, another advantage of this framework is parallelization of the assembly and element result computations without requiring additional communication between processors.

As a conclusion, this study presented the development of a parallel substructure based solution framework designed for the efficient linear static analysis of large structures having multiple loading conditions. The framework can be applied to models which contain a mixture of element types and to models which exceed the memory capacity of the processors.

7.2 Future Work

There are many extensions and improvements which can be made to the presented framework that would increase its efficiency and functionality. Some of these are discussed below:

- **New equation numbering algorithm:** Currently, the internal and interface equations are numbered independently. As a result, the column heights of the interface equations are much higher than the column heights of the internal equations which not only increases the number of operations for condensation but also the storage requirements. Moreover, the substructure condensation times become more sensitive to any modifications in their shapes during the workload balancing iterations due a lack of control on the interface column heights. A new equation numbering algorithm which optimizes the equation numbering by considering the constraint of having interface equations assembled at the end of the matrix, would not only decrease the condensation

257

time but also increases the amount of improvement obtained during the workload balancing step.

- **Automatic processor assignment for interface solution:** A sub-program that will examine the computational and communication properties of a cluster and determine the optimum number of processors to solve according to the size of the interface problem will improve the efficiency of the current framework. For example, if a cluster has low communication speed with fast computation speed, utilizing a large number of processors may slow down the interface solution. For such cases, the condensations can be performed by utilizing the maximum available processors but the interface equations can be solved with a smaller number of processors which will maximize the speed-up for the interface solution.

- **Determining the optimum number of processors for solution:** The scalability of this framework depends on the size of the model. It was observed that decreasing the size of the substructures after a point did not decrease the condensation time but increased the interface solution time considerably. Thus, an algorithm which determines the optimum number of substructures considering the size of the structure and the computation and communication properties of a cluster will improve the performance of this framework significantly.

- **Heterogeneous computing environment:** Civil engineering design offices may have computers that have varying computational speeds. The solution framework designed for such parallel environments would be very useful. The

258

current structure of the presented framework allows the solution algorithms to work with such systems with minor modifications. For example, during the workload balancing iterations, a scaling variable could be added to imbalance factor calculation, Equation 4.1, to represent the relative computational speeds of computers. This way, the resulting substructures will have the condensation time ratios closer to their processor's computational speed ratios. Moreover, the rows of the interface stiffness matrix can be distributed to the processors according to their computational speeds. In other words, faster processors would store more rows. For both cases, more detailed examination of the computational properties of computers having different hardware must be performed.

- **Rigid body constraint:** One of the important civil engineering structures is the concrete high-rise building. Generally, the slabs and walls of such buildings are modeled with shell elements whereas the beams and the columns are modeled with frame elements. For most of the cases, the floors are assumed to be rigid. This rigidity condition is applied with the constraining equations that describe the relationship between the various degrees of freedom of the stiffness matrix. They are applied to the global stiffness matrix by three different methods: transformation, Lagrange multipliers and penalty function. A parallel substructuring method, on the other hand, performs partial assembly on different processors. In other words, each processor has only a part of the global stiffness matrix. Thus, efficient implementation of the application of

259

rigid body constraints for the substructure based parallel solution would improve the framework's functionality.

- **Solution of problems having repetitive right hand sides:** This framework can also be utilized to solve problems having repetitive right hand sides such as linear time history analysis with implicit integration or non-linear analysis based on a modified Newton-Raphson method. In this case, the substructure level and the interface stiffness matrices are factorized once, and the load factorizations are performed for each right hand side vector. Thus, the factorizations and forward substitutions can not be performed together which causes a significant communication overhead during the interface solution because the lower triangular coefficients must be distributed among the processors for each forward substitution. Even if the all the lower triangular coefficients are stored at each processor's local memory, there will be a start-up latency cost. Thus, any approach that reduces the communication cost for the interface load factorizations will improve the performance of the such a framework.

APPENDIX A

A.1 Example Problems

Six different structural models were utilized in this study. The first model is a square mesh model which is mostly utilized for illustrative purposes. The second and third models are 3D models. They are generated from actual structural models. The last three models are structures that are either being designed or constructed.

A.1.1 2D Square Mesh

The 2D Square Mesh model is composed of 25,600 quadrilateral shell elements as shown in Figure A.1.

# Nodes	# Equations	# Members	# Elements
25,921	155,526	-	25,600

Figure A.1 2D Square Mesh Model

A.1.2 Half Disk

Half-disk model is composed of 28,128 eight node brick elements as shown in Figure A.2.

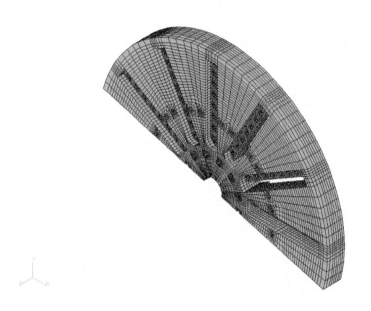

# Nodes	# Equations	# Members	# Elements
36,773	110,319	-	28,128

Figure A.2 Half-disk Model

A.1.3 Bridge Deck

The Bridge Deck model is a highway bridge model having 24,200 eight node brick elements. The bridge is being constructed on I-20 near, Atlanta, Georgia. The size of the model was increased for research purposes.

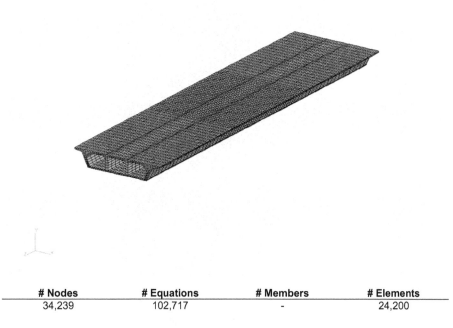

# Nodes	# Equations	# Members	# Elements
34,239	102,717	-	24,200

Figure A.3 Bridge Deck Model

A.1.4 High-rise Building I

The High-rise Building I model is composed of 1,221 frame members, 61,307 quadrilateral and triangular shell elements. The building has 26 stories and is being constructed in Destin, Florida. It has 16 shear walls and the slab geometry remains almost the same at every floor.

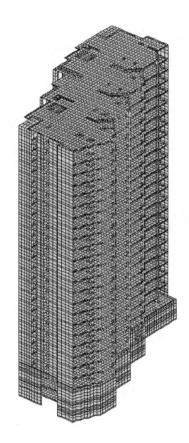

# Nodes	# Equations	# Members	# Elements
54,751	328,506	1,221	61,307

Figure A.4 High-rise Building I Model

A.1.5 High-rise Building II

The High-rise Building II model is composed of 6,001 frame members and 54,450 quadrilateral and triangular shell elements. The building has both hotel and residential spaces and will be built in Atlanta, Georgia. It has 27 stories. The lower levels are highly irregular with the floor plan changing at each level and with large openings in the floors.

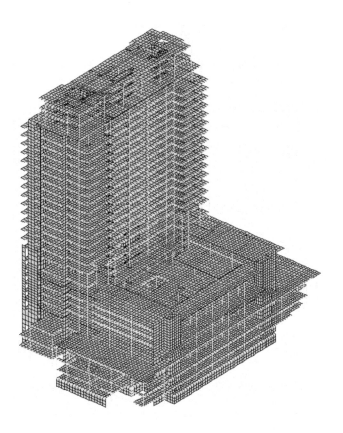

# Nodes	# Equations	# Members	# Elements
54,773	328,638	6,001	54,450

Figure A.5 High-rise Building II Model

A.1.6 Nuclear Waste Plant

The Nuclear Waste Plant model is composed of 2,811 frame members and 43,776 quadrilateral shell elements. There are many rooms inside the bottom box which increase the solution skyline significantly.

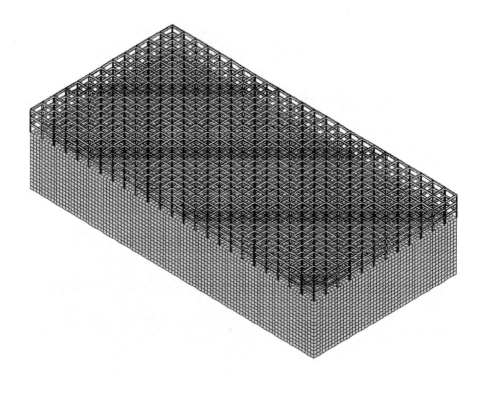

# Nodes	# Equations	# Members	# Elements
39,440	236,640	2,811	43,776

Figure A.6 Nuclear Waste Plant Model

APPENDIX B

B.1 Unified Modeling Language

B.1.1 History

During the nineties, many different methodologies each with their own set of notations were introduced [133]. Among these methods, three of them became the most popular ones: OMT [135], Booch [130], and OOSE [132]. Each method had its own strong points and weaknesses. For example, Booch [130] was strong in design and weaker in analysis; however OMT [135] was strong in analysis and weaker in design [133].

Booch [129] wrote his second book in 1995, Rumbaugh developed OMT-2 [134] which incorporated many of the good points of other methods. Although the methods became more similar to each other, each of them had its own unique notations. This period of time was remembered as "Method Wars" [133].

Finally, the methods were joined under the "Unified Modeling Language (UML)" which represented the unification of the Booch [130], OMT[135] and OOSE [132] notations as well as the best ideas from other methodologists. Due to this unification, UML became the standard in the domain of object-oriented analysis and design.

B.1.2 Structure

The UML is a modeling language for specifying, visualizing constructing and documenting the artifacts of a system. It is defined within a conceptual framework that consists of four distinct levels of abstraction.

- **The meta-metamodel layer:** Consists of the concept of a "Thing", representing anything that may be defined.

- **Metamodel layer:** Consists of elements that constitute the UML including concepts from the object-oriented and component oriented paradigms.

- **Model layer:** Consists of UML models. At this level modeling of problems, solutions or systems occur. Model in this layer are called class or type models.

- **User model layer:** Consists of elements that exemplify UML models. Models in this layer are called object or instance model.

UML provides different models in order to capture the structural and the behavioral features of systems. It has diagrams for these models in order to depict knowledge in a communicable form [136]:

- The User Model View
    - → Use Case Diagrams
- The Structural Model View
    - → Class Diagrams
    - → Object Diagrams
- The Behavioral Model View
    - → Sequence Diagrams
    - → Collaboration Diagrams
    - → State Diagrams
    - → Activity Diagrams

268

- Implementation Model View

  → Component Diagrams

- Environment Model View

  → Deployment Diagrams

Only the classes and the class diagrams will be discussed in this section since they are utilized to describe the database structure of the parallel solution program.

B.1.2.1 Structural Model

<u>Class</u>

A class is a description of a group of objects with common properties, common behavior, common relationships to other objects and common semantics. There is not a unique method for finding classes in a system. UML proposes to look for three types of classes during the discovery process: Entity, boundary and control classes.

- *Entity Classes:* They model information and associated behavior that is generally long lived. They may reflect a real world entity, or it may be needed to perform tasks internal to the system. They are independent of their surroundings, and the application. They may be used in more than one application.

- *Boundary Classes:* They handle the communication between the system surroundings and the inside of the system. They can provide the interface to a user or another system.

- *Control Classes:* They model sequencing behavior specific to one or more use cases. They coordinate the events needed to realize the behavior specified in the use case. They are typically the application dependent classes.

Class Diagrams:

The class diagrams are utilized to provide a picture or a view of some or all of the classes in the model. Classes are connected to each other with relationships. UML differentiates the following relationships.

- *Association Relationship:* It is a bi-directional semantic connection between classes. An association between classes means that there is a link between objects in the associated class.

  Symbol: ─────────────

- *Aggregation Relationship:* It is a specialized form of association in which a whole is related to its parts. Aggregation is known as a "part of" relationship.

  Symbol: ───────────◇

- *Multiplicity:* It defines the number of object that is linked to one another.

  Symbol: 1..* : at least one instance

         1: exactly one instance

        0..* : no limit on the number of instances

- *Generalization Relationship:* An inheritance link indicating one class is a superclass of the other.

  Symbol: ───────────▷

A class embodies a set of responsibilities that define the behavior of the objects in the class. The responsibilities are carried out by the operations defined for the class. An operation does only one thing.

The structure of an object is described by the attributes of the class. Each attribute is a data definition held by objects of the class. Objects defined for the class have a value for every attribute of the class.

The class diagrams show each objects structure and behavior and the relations between them.

B.2 Database Structure of the Parallel Solution Program

The parallel solution program was developed by using both C++ and FORTRAN programming languages. The MPICH [115] was utilized for parallelism. For shared memory architectures, the WIN32 thread libraries were utilized. The program has an object-oriented database structure. The class diagrams and their relationships of important database elements are presented in this section by using Unified Modeling Language (UML) [136].

The very top level of the database is presented in Figure B.1. An abstract class named 'System' is created in order to store the entire information about a project. A system can be composed of one or more structures. The 'Structure' class in a system represents an actual independent structural model having nodes, elements, loads etc. Each structure is analyzed separately. Furthermore, a structure can be composed of mechanically dependent substructures. A 'Substructure' class is created for that purpose. Each substructure has its own elements, loads and nodes. The substructures can be utilized in both serial and parallel computations.

The material definitions are stored in the structure's database. This way, the same material definition can be used by different substructures. Furthermore, more than one solution algorithm can be added to a structure's database so that different analyses can be

performed on the same structure consecutively.

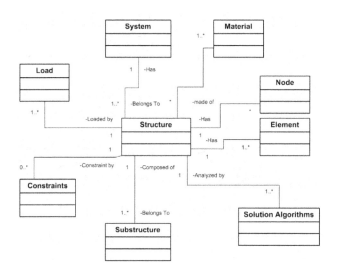

Figure B.1 The Class Diagram for the Global System

The program database is composed of three main libraries: mechanical library, matrix

library and solution algorithms library. The mechanical library defines the relationships

between the real mechanical entities like nodes, material, element, and loads. Its structure

is presented in Figure B.2.

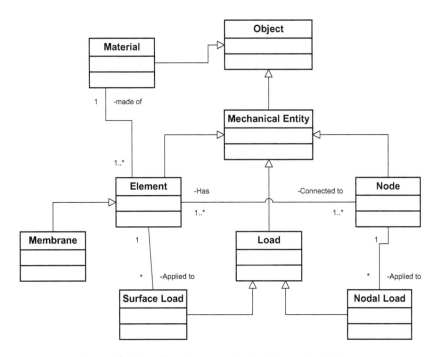

Figure B.2 The Class Diagram for the Mechanical Library

Every class is assumed to be inherited from the abstract 'Object' class. Likewise, all mechanical entities are assumed to be inherited from an abstract 'Mechanical Entity' class. 'Element' class is an abstract class for all types of finite elements, frame members, springs etc. Each element has to be connected to one or more nodes. Similarly, each node should have one or more elements attached to it. 'Load' class is created as an abstract class for all types of loadings.

The solution algorithms library aims to collect all kinds of analysis methods from static to dynamic, linear to non-linear and serial to parallel. Each solution algorithm class needs to be a child class of the abstract 'Solution Algorithms' Class. The current program has various types of solution algorithm classes. The linear static solution algorithm classes are as follows:

➢ Substructure Based Solution Methods

- Parallel Linear Solution with Multiple Loading Conditions (MPI)
  - o In-core Condensation
  - o Out-of-core Condensation
- Serial Linear Solution with Multiple Loading Conditions
  - o In-core Condensation
  - o Out-of-core Condensation
- Out-of core, Serial Block Solution Methods
- Shared, active column, LU Decomposition [40] (Thread)
- Serial, active column, block LU Decomposition

➢ In-core Solution Methods

- Serial, active column, LU Decomposition
- Serial, constant band, LU Decomposition

The other important database element is the 'Substructure' class. It is the key element of the substructure based solution algorithms. The class relationship diagram of the 'Substructure Class' is given in Figure B.3.

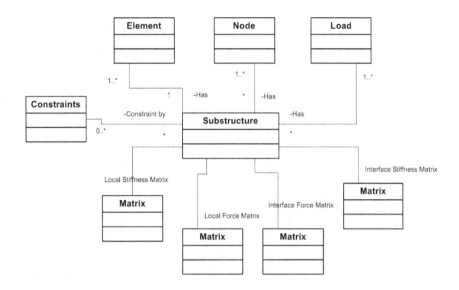

Figure B.3 The Class Diagram for the Substructure Class

Each substructure has a reserved space for its own local stiffness matrix and force vectors. During the solution, the solution algorithm class uses this space to assemble the local stiffness matrix and force vectors. The interface stiffness and force matrix data space are used to keep the data of the assigned rows of the interface stiffness matrix during the parallel solution of the interface equations.

The connectivity information needs to be stored in order to keep track of the shared nodes when a structure is partitioned into more than one substructure. The connectivity information is utilized during condensation to distinguish the interface nodes from the internal nodes and during interface solution while preparing the data distribution scheme of the interface equations. For that purpose, two new classes are created to hold the shared node information among the substructures. The first one is called 'Node Tie' class which keeps the pointers of two overlapping nodes. The second one, 'Substructure Tie', is actually a collection of node ties which defines the connection between two substructures. A representation of the internal and interface nodes, substructure and node ties is shown in Figure B.4.

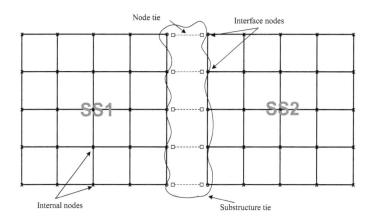

Figure B.4 The Substructure Representation of a Structure

The 'Parallel Substructure Solver' abstract class was created to facilitate the addition of a new parallel substructure solution algorithm. Moreover, it allows changing the data distribution scheme and contains general subroutines that will be used by all of the solution algorithms. These subroutines use 'DistributionList' and 'DofMapper' classes. This way, even if a new algorithm needs a different distribution or different numbering scheme, it can still use these subroutines. Moreover, another abstract class for the data distribution algorithms was created. All the data distribution algorithms must be its child classes and use 'DofMapper' and 'DistributionList' classes. Therefore, during the data distribution step, the selected data distribution algorithm's job is to prepare 'DofMapper' and 'DistributionList' classes. The rest is handled by the solution algorithms. The Data

276

Distribution Algorithm is designed as a part of the Parallel Substructure Solver class. If,

in the future, the method for data distribution needs to be changed, a new child class of

the 'Data Distribution Algorithm' must be created and added to the Parallel Substructure

Class. The class diagram of Parallel Substructure solver is given in Figure B.5

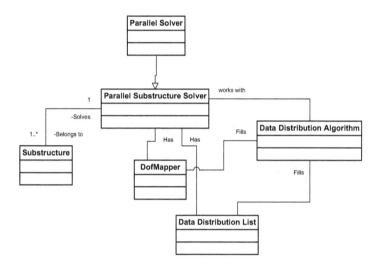

Figure B.5 The Class Diagram for Parallel Substructure Solver Class

REFERENCES

[1] S. T. Barnard, H. D. Simon, Fast Multilevel Implementation of Recursive Spectral Bisection for Partitioning Unstructured Problems, Concurrency: Practice and Experience 6 (1994) 101-117

[2] M.J. Berger, S.H. Bokhari, A Partitioning Strategy for Non-uniform Problems on Multiprocessors, IEEE Trans. Computers, C 36 (1987) 570-580

[3] R. Biswas, L. Oliker, Experiments with Repartitioning and Load Balancing of Adaptive Meshes, Grid Generation and Adaptive Algorithms, IMA Volumes in Mathematics and its Applications, Vol. 113, (1999), 89-112.

[4] U.V. Catalyurek, C. Aykanat, Hypergraph Partitioning Based Decomposition for Parallel Sparse Matrix-Vector Multiplication, IEEE Trans. Parallel Distrib. Syst. 10 (1999) 673-693

[5] G. Cybenko, Dynamic Load Balancing for Distributed Memory Multiprocessors, J. Parallel Distrib. Comput. 7 (1989) 279-301

[6] D.M. Day, M.K. Bhardwaj, G.M. Reese, J.S. Peery, Mechanism Free Domain Decomposition, Comput. Methods Appl. Mech. Engrg. 192 (2003) 763-776

[7] R.V. Driessche, D. Roose, An Improved Spectral Bisection Algorithm and Its Application to Dynamic Load Balancing, Parallel Comput. 21 (1995) 29-48

[8] C. Farhat, N. Maman, G. W. Brown, Mesh Partitioning for Implicit Computations via Iterative Domain Decomposition: Impact and Optimization of the Subdomain Aspect Ratio, Int. J. Numer. Methods Eng., 38 (1995) 989-1000

[9] C. Farhat, A Simple and Efficient Automatic FEM Domain Decomposer, Comput. Struct. Vol. 28 No.5 (1988) 579-602

[10] C. M. Fiduccia, R. M. Mattheyses, A Linear Time Heuristics for Improving Network Partitions, 19th IEEE Design Automation Conference, IEEE, 1982, 175-181

[11] M. Fiedler, Algebraic Connectivity of Graphs, Czechoslovak Math. J., 23 (1973) 298-305

[12] J.E. Flaherty, R.M. Loy, M.S. Shephard, B.K. Szymanski, J.D. Teresco, L.H. Ziantz, Adaptive Local Refinement with Octree Load Balancing for the Parallel Solution of Three Dimensional Conservation Laws, J. Parallel Distrib. Comput. 47 (1998) 139-152

[13] M. Garey, D. Johnson, L. Stockmeyer, Some Simplified NP-complete Graph Problems, Theoretical Computer Science, 1 (1976), pp. 237-267

[14] N.E. Gibbs, W.G Poole JR., P.K. Stockmeyer, An Algorithm for Reducing the Bandwidth and Profile of a Sparse Matrix, SIAM J Numer. Anal 13, 2, April, (1976) 236-250

[15] N.E. Gibbs, Algorithm 509: A Hybrid Profile Reduction Algorithm, ACM Trans. On Math Software, 2, (1976) 378-387

[16] B. Hendrickson, K. Devine, Dynamic Load Balancing in Computational Mechanics, Comput. Methods Appl. Engrg. 184 (2000) 485-500

[17] B. Hendrickson, T.G. Kolda, Graph Partitioning Models for Parallel Computing, Parallel Comput. 26 (2000) 1519-1534

[18] B. Hendrickson, T.G. Kolda, Partitioning Non-square and Non-symmetric Matrices for Parallel Processing, SIAM J. Sci. Comput. 21 (2000) 2048 – 2072

[19] B. Hendrickson, Load Balancing Fictions, Falsehoods and Fallacies, Appl. Math. Model. 25 (2000) 99-108

[20] B. Hendrickson, R. Leland, R.V. Driessche, Skewed Graph Partitioning, Proceedings of the Eighth SIAM Conference Parallel Processing for Scientific Computing, SIAM, 1997

[21] B. Hendrickson, R. Leland, A Multilevel Algorithm for Partitioning Graphs, Technical Report SAND93-1301, Sandia National Laboratories, (1993)

[22] B. Hendrickson and R. Leland, An Improved Spectral Partitioning Algorithm for Mapping Parallel Computations, Sandia Report, SAND92-1460, Category UC-405, Sandia National Laboratories, Albuquerque, NM 87185, (1992)

[23] S.H. Hsieh, Y.S. Yang, P.L. Tsai, Improved Mesh Partitioning for Parallel Substructure Finite Element Computations, Proceedings of the 7th East Asia-Pacific Conference on Structural Engineering and Construction, (1999), 123-128

[24] S.H. Hsieh, G.H. Paulino, J. F. Abel, Recursive Spectral Algorithms for Automatic Domain Partitioning in Parallel Finite Element Analysis, Comput. Methods Appl. Mech. Engrg. 121 (1995) 137-162

[25] Y.F. Hu, R.J. Blake, D.R. Emerson, An Optimal Migration Algorithm for Dynamic Load Balancing, Concurrency: Pract. Exper., Vol. 10 No.6 (1998) 467-483

[26] M.T. Jones, P.E. Plassman, Computational Results for Parallel Unstructured Mesh Computations, Comput. Systems Engr. Vol.5 No.4-6 (1994) 297-309

[27] B.W. Kernighan, S. Lin, An Efficient Heuristic Procedure for Partitioning Graphs, BELL Syst. Tech. J., Feb. (1970) 291-307

[28] G. Karypis, K. Schloegel, V. Kumar, PARMETIS: A Parallel Graph Partitioning and Sparse Matrix Ordering Library, version 3.1, (2003)

[29] G. Karypis, V. Kumar, Multilevel Algorithms for Multi-constraint Graph Partitioning, Technical Report, TR 98-019, Department of Computer Science, University of Minnesota, (1998)

[30] G. Karypis, V. Kumar, METIS: a Software Package for Partitioning Unstructured Graphs, Partitioning Meshes, and Computing Fill-reducing Orderings of Sparse Matrices, version 4.0, (1998)

[31] G. Karypis, V. Kumar, METIS: A Software Package for Partitioning Unstructured Graphs, Partitioning Meshes, and Computing Fill-reducing Orderings of Sparse Matrices, ver. 4.0, (1998) (available at http://www-users.cs.umn.edu/~karypis/metis/ , 03/01/2005)

[32] G. Karypis, V. Kumar, Multilevel k-way Partitioning Scheme for Irregular Graphs, Technical Report, TR 95-064, Department of Computer Science, University of Minnesota, (1995)

[33] G. Karypis, V. Kumar, A Fast and High Quality Multilevel Scheme for Partitioning Irregular Graphs, Technical Report, TR 95-035, Department of Computer Science, University of Minnesota, (1995)

[34] J.G. Malone, Automated Mesh Decomposition and Concurrent Finite Element Analysis for Hypercube Multiprocessor Computers, Comput. Methods Appl. Mech. Engrg. 133 (1996) 25-45

[35] L. Oliker, R. Biswas, PLUM: Parallel Load Balancing for Adaptive Unstructured Meshes, Journal of Parallel Dist. Comp., 52, 2, (1998), 150-177

[36] C.W. Ou, S. Ranka, Parallel Incremental Graph Partitioning using Linear Programming. Technical report, Syracuse University, Syracuse, NY, (1992)

[37] K. Schloegel, G. Karypis, V. Kumar, Parallel Multilevel Algorithms for Multi-constraint Graph Partitioning, Technical Report TR 99-031, Department of Computer Science, University of Minnesota, (1999)

[38] K. Schloegel, G. Karypis, V. Kumar, Parallel Multilevel Algorithms for Multi-Constraint Graph Partitioning, Technical Report TR 99-031, Department of Computer Science, University of Minnesota, (1999)

[39] K. Schloegel, G. Karypis, V. Kumar, Wavefront Diffusion and LMSR: Algorithms for Dynamic Repartitioning of Adaptive Meshes, Technical Report TR 98-034, Department of Computer Science, University of Minnesota, (1998)

[40] K. Schloegel, G. Karypis, V. Kumar, R. Biswas, L. Oliker, A Performance Study of Diffusive vs. Remapped Load-Balancing Schemes, Technical Report TR 98-018, Department of Computer Science, University of Minnesota, (1998)

[41] K. Schloegel, G. Karypis, V. Kumar, Multilevel Diffusion Schemes for Repartitioning of Adaptive Meshes, Technical Report TR 97-013, Department of Computer Science, University of Minnesota, (1997)

[42] K. Schloegel, G. Karypis, V. Kumar, Parallel Multilevel Diffusion Algorithms for Repartitioning of Adaptive Meshes, Technical Report TR 97-014, Department of Computer Science, University of Minnesota, (1997)

[43] H.D. Simon, A. Sohn, R. Biswas, HARP: A Dynamic Spectral Partitioner, J. Parallel Distrib. Comput. 50 (1998) 83-103

[44] H.D. Simon, Partitioning of Unstructured Problems for Parallel Processing, Proceedings of: Parallel Methods on Large Scale Structural Analysis and Physics Appl., Pergamon, New York, 1991

[45] F. Pellegrini, Graph Partitioning Based Methods and Tools for Scientific Computing, Parallel Comput. 23 (1997) 153-164

[46] F. Pellegrini, J. Roman, SCOTCH: A Software Package for Static Mapping by Dual Recursive Bipartitioning of Process and Architecture Graphs, Proc. HPCN'96, Brussels, LNCS 1067, April 1996 493-498

[47] A.Pothen, H.D. Simon, K. Liou, Partitioning sparse matrices with eigenvectors of graphs, SIAM J. Matrix Anal. Appl., Vol. 11 No. 3 (1990) 430-452

[48] J.D. Teresco, M.W. Beall, J.E. Flaherty, M.S Shephard, A Hierarchical Partition Model for Adaptive Finite Element Computation, Comput. Methods Appl. Mech. Engrg. 184 (2000) 269-285

[49] B.H.V. Topping, P. Ivanyi, Partitioning of Tall Buildings using Bubble Graph Representation, J. Comput. Civil Engrg. Vol. 15 No.3 (2001) 178-183

[50] D. Vanderstraeten, C. Farhat, P. S. Chen, R. Keunings, O. Ozone, A Retrofit Based Methodology for the Fast Generation and Optimization of Large-Scale Mesh Partitions: Beyond the Minimum Interface Size Criterion, Comput. Methods Appl. Mech. Engrg. 133 (1996) 25-45

[51] D. Vanderstraeten, R. Keunings, Optimized Partitioning of Unstructured Finite Element Meshes, Int. J. Numer. Methods Eng., 38 (1995) 433-450

[52] C. Walshaw, M. Cross, Parallel Optimization Algorithms for Multilevel Mesh Partitioning, Parallel Comput. 26 (2000) 1635-1660

[53] C. Walshaw, M. Cross, M. G. Everett, Dynamic Load-balancing for Parallel Adaptive Unstructured Meshes. Parallel Processing for Scientific Computing, (1997)

[54] C. Walshaw, M. Berzins, Dynamic Load-balancing for PDE Solvers on Adaptive Unstructured Meshes, Concurrency: Pract. Exp. 7 (1995) 17-28

[55] P.R. Amestoy, I.S. Duff, J.-Y. L'Excellent, Multifrontal Parallel Distributed Symmetric and Unsymmetric Solvers, Comput. Methods Appl. Engrg. 184 (2000) 501-520

[56] P.R. Amestoy, T. A. Davis, I. S. Duff, An Approximate Minimum Degree Ordering Algorithm, SIAM J. Matrix Analy. Appl. 17 (1996) 886-905

[57] C. Ashcraft, S. Eisenstat, J. Liu, A Fan-in Algorithm for Distributed Sparse Numerical Factorization, SIAM J. Sci. Stat. Comput. 11 (1990) 593-599

[58] C. Ashcraft, The Fan-both Family of Column-based Distributed Cholesky Factorization Algorithms, Graph Theory and Sparse Matrix Computation, The IMA Volumes in Mathematics and its Applications, Springer, Berlin, New York, 56 (1993) 159-190

[59] C. Ashcraft, R.G Grimes, B. W. Peyton, H. D. Simon, Progress in Sparse Matrix Methods for Large Linear Systems on Vector Supercomputers, Int. J. Supercomput. Appl. 1 (4) (1987) 10-30

[60] J.W. Baugh Jr., S.K. Sharma, Evaluation of Distributed Finite Element Algorithms on a Workstation Network, Engrg. Comput. 10 (1994) 45-62

[61] M. Bhardwaj, D. Day, C. Farhat, M. Lesoinne, K. Pierson, D. Rixen, Application of the FETI Method to ASCI Problems, Scalability Results on 1000 Processors and Discussion of Highly Heterogeneous Problems, Int. J. Numer. Meth. Engng. 47, (2000) 513-535

[62] S. Bitzarakis, M. Papadrakakis, A. Kotsopulos, Parallel Solution Technique in Computational Structural Mechanics, Comput. Methods Appl. Mech. Engrg. 148 (1997) 75-104

[63] K. N. Chiang, R. E. Fulton, Concepts and Implementation of Parallel Finite Element Analysis, Comput. Struct., Vol. 36, No. 6 (1990) 1039-1046

[64] S. C. Chuang, R. E. Fulton, Decomposition of Sparse Matrices on Parallel Computers, Comput. Systems. Engng. Vol. 3 No. 1-4 (1992) 357-363

[65] J.M. Conroy, S.G. Kratzer, R.F. Lucas, Data-parallel Sparse Matrix Factorization, in: Proc. 5th SIAM Conference on Linear Algebra, SIAM, Philadelphia, PA, (1994) 377

[66] R. D. Cook, D. S. Malkus, M. E. Plesha, Concepts and Applications of Finite Element Analysis, John Wiley and Sons Inc, (1989)

[67] I. S. Duff, H. A. van der Vorst, Developments and Trends in the Parallel Solution of Linear Systems, Parallel Comput. 25 (1999) 1931-1970

[68] I. S. Duff, J.A. Scott, The Use of Multiple Fronts in Gaussian Elimination, Proc. Fifth SIAM Conf. on Applied Linear Algebra, SIAM, Philadelphia, PA, (1994) 567-571

[69] I. S. Duff, The Impact of High Performance Computing in the Solution of Linear Systems: Trends and Problems, J. Comput. Appl. Math. 123 (2000) 515-530

[70] I. S. Duff, A Review of Frontal Methods for Solving Linear Systems, Comput. Phys. Comm. 97 (1996) 45-52

[71] Y. Escaig, G. Touzot, M. Vayssade, Parallelization of a Multilevel Domain Decomposition Method, Comput. Systems Engrg. Vol.5 No.3 (1994) 253-263

[72] G. P. Nikishkov, A. Makinouchi, G. Yagawa, S. Yoshimura, Performance Study of the Domain Decomposition Methods with Direct Equation Solver for Parallel Finite Element Analysis, Comput. Mech. 19 (1996) 84-93

[73] C. Farhat, K. Pierson, M. Lesoinne, The Second Generation FETI Methods and Their Application to the Parallel Solution of Large-scale Linear and Geometrically Non-linear Structural Analysis Problems, Comput. Methods Appl. Mech. Engrg. 184, (2000) 333-374

[74] C. Farhat, P.S. Chen, F. Risler, F. X. Rous, A Unified Framework for Accelerating the Convergence of Iterative Substructuring Methods with Lagrange Multipliers, Int. J. Numer. Methods Engrg. 42 (1998) 257-288

[75] C. Farhat, P. S. Chen, P. Stern, Towards the Ultimate Iterative Substructuring Method: Combined Numerical and Parallel Scalability and Multiple Load Cases, Vol. 5, No. 4-6 (1994) 337-350

[76] C. Farhat, L. Crivelli, F. X. Roux, Extending Substructure Based Iterative Solvers to Multiple Load and Repeated Analyses, Comput. Methods Appl. Mech. Engrg. 117 (1994) 195-209

[77] C. Farhat, F.-X. Roux, A Method of Finite Element Tearing and Interconnecting and its Parallel Solution Algorithm, Int. J. Numer. Methods Engrg. 32 (1991) 1205-1227

[78] C. Farhat, Redesigning the Skyline Solver for Parallel/Vector Supercomputers, Int. J. High Speed Comput., Vol.2 No.3 (1990) 223-238

[79] C. Farhat, E. Wilson, A Parallel Active Column Equation Solver, Comput. Struct., Vol. 28 No.2 (1988) 289-304

[80] C. Farhat, E. Wilson, A New Finite Element Concurrent Computer Program Architecture, Int. J, Num. Methods. Engrg., Vol.24 (1987) 1771-1792

[81] C. Farhat, E. Wilson, G. Powell, Solution of Finite Element Systems on Concurrent Processing Computers, Engineering with Computers, 2 (1987) 157-165

[82] R. E. Fulton, P. S. Su, Parallel Substructure Approach for Massively Parallel Computers, Comput. Engng. Vol. 2 (1992) 75-82

[83] A. Guermouche, J.-Y. L'Excellent, Impact of Reordering on the Memory of a Multifrontal Solver, Parallel Comput., 29 (2003) 1191-1218

[84] A. George, M. Health, J. Liu, E. Ng, Solution of Sparse Positive Definite Systems on a Hypercube, J. Comp. Appl. Math, 27 (1989) 129-156

[85] A. S. Gullerud, R. H. Dodds Jr., MPI-based Implementation of a PCG Solver Using an EBE Architecture and Preconditioner for Implicit 3-D Finite Element Analysis, Comput. & Struct. 79 (2001) 553-575

[86] T. Y. Han, J. F. Abel, Substructure Condensation Using Modified Decomposition, Int. J. Num. Methods. Engng, Vol.20 (1984) 1959-1964

[87] M. Heath, P. Raghavan, Performance of a Fully Parallel Sparse Solver, Proc. Scalable High Performance Computing Conference, IEEE Computer Society Press, (1994) 334-341

[88] M. Heath, P. Raghavan, Distributed Solution of Sparse Linear Systems, Proc. Scalable Parallel Libraries Conference, IEEE Computer society Press, (1994) 114-122

[89] S. H. Hsieh, S. Modak, E. D. Sotelino, Object-oriented Parallel Programming Tools for Structural Engineering Applications, Comput. Systems Engng., Vol. 6, No. 6, (1995) 533-548

[90] B. M. Irons, A Frontal Solution Program for Finite Element Analysis, Int. Jour. Num. Meth. Engng., Vol. 2, (1970), 5-32

[91] O. Kurc, K.M. Will, A Parallel Active Column Matrix Solution Algorithm for Systems with Multiple Loading Cases, Proceedings of the Ninth International Conference on Computing in Civil and Building Engineering, 2002

[92] J. Mackerle, Parallel Finite Element and Boundary Element Analysis: Theory and Applications – A bibliography (1997-1999), Fin. El. Analy. Des., 35 (2000) 283-296

[93] S. Modak, E. D. Sotelino, S. H. Hsieh, A Parallel Matrix Class Library in C++ for Computational Mechanics Applications, Microcomputer in Civil Engng., 12 (1997) 83-99

[94] C. S. R. Murthy, K. N. B. Murthy, S. Aluru, New Parallel Algorithms For Direct Solution of Linear Equations, Wiley, New York (2000)

[95] M. Papadrakakis, S. Smerou, A New Implementation of the Lanczos Method in Linear Problems, Int. J. Numer. Meth. Engrg. 29 (1990) 141-159

[96] Parasol Project, http://www.parallab.uib.no/parasol/ (01/03/2005)

[97] K. C. Park, M. R. Justino Jr., C. A. Felippa, An Algebraically Partitioned FETI Method for Parallel Structural Analysis: Algorithm Description, Int J. Numer. Methods Engng 40 (1997) 2717-2737

[98] J. Schulze, Towards a Tighter Coupling of Bottom-up and Top-down Sparse Matrix Ordering Methods, BIT 41 (4) (2001) 800-841

[99] R. Schreiber, Scalability of Sparse Direct Solvers, Graph Theory and Sparse Matrix Computation, The IMA Volumes in Mathematics and its Applications, Springer, Berlin, New York, 56 (1993) 191-209

[100] S. Y. Synn, R. E. Fulton, Practical Strategy for Concurrent Substructure Analysis, Comput. Struct. Vol. 54 No. 5 (1995) 939-944

[101] E. L. Wilson, H. H. Dovey, Solution or Reduction of Equilibrium Equations for Large Complex Structural Systems, Adv. Eng. Soft. Vol. 1 No.1 (1978) 19-25

[102] Y. S. Yang, S. H. Hsieh, Iterative Mesh Partitioning Optimization for Parallel Nonlinear Dynamic Finite Element Analysis with Direct Substructuring, Comput. Mechs., 28 (2002) 456-468

[103] ASCI Cluster, http://www.cacr.caltech.edu/beowulf/ (01/03/2005)

[104] G. H. Barnes et al. The Iliac IV Computer, IEEE Trans. Computers, Vol. c-17, Aug. (1968) 746-757

[105] R. G. Brown, Engineering a beowulf-style computer cluster, Physics Department, Duke University, (2004)

[106] T.L. Casavant, S. A. Fineberg, M. L. Roderick, B. H. Pease, Massively Parallel Architectures

[107] C. Chakrabarti, J. Demmel, K. Yelick, Modeling the benefits of mixed data and task parallelism, Technical Report UT-CS-95-289, Univ. Tennessee, Knoxville

[108] D. Cheriton, The v distributed system. Communications of the ACM, March 31, 3 March (1988)

[109] K. Dowd, C. Severance, High Performance Computing, O'Reilly (1993)

[110] M. J. Flynn, Some computer organizations and their effectiveness, IEEE Transactions on Computing, C-21, (1972) 948-960

[111] H. Forsdick, R.E.Schantz, and R. H. Thomas, Operating systems for computer

Networks, Computer 11, 1 January (1978), 48-57

[112] IBM Blue Gene/L, http://www.research.ibm.com/bluegene (01/03/2005)

[113] A. A. Khokhar, C.-L. Wang, M. Shaaban, and V. Prasanna, Heterogeneous computing: Challenges and opportunities. IEEE Computer Magazine, Special Issue on Heterogeneous Processing 26, 6 June (1993)

[114] B. Meyer, Object-oriented Software Construction, (1997)

[115] MPICH Library, http://www-unix.mcs.anl.gov/mpi/mpich/ (01/03/2005)

[116] Nakashima, M., Kato, H., and Takaoka, E., "Development of Real-Time Pseudo Dynamic Testing." Earthquake Engineering and Structural Dynamics, 21. (1999) 79–92.

[117] NEESgrid, http://www.neesgrid.org (01/03/2005)

[118] NEC Tx7, http://www.clustervision.com/tx7.html (01/03/2005)

[119] Top 500 Supercomputer sites, www.top500.org (01/03/2005)

[120] Ashcraft, C., Grimes, R.G. (1999). "SPOOLES: An Object-Oriented Sparse Matrix Library", Proceedings of 1999 SIAM Conference on Parallel Processing for Scientific Computing, March 22-27. (available at http://www.netlib.org/linalg/spooles, o1/03/2005)

[121] The Globus Project, www.globus.org (01/03/2005)

[122] GTSTRUDL, A Structural Analysis and Design Software, CASE Center, School of Environmental and Civil Engineering, Georgia Institute of Technology, Atlanta, GA, www.gtstrudl.gatech.edu (01/03/2005)

[123] PACI Genie Work, http://www.npaci.edu/ (01/03/2005)

[124] L. Pearlman, C. Kesselman, S. Gullapalli, B.F. Spencer, Jr., J. Futrelle, K. Ricker, I. Foster, P. Hubbard and C. Severance, Distributed Hybrid Earthquake Engineering Experiments: Experiences with a Ground-Shaking Grid Application., Proceedings of the 13th IEEE Symposium on High Performance Distributed Computing, June (2004)

[125] J.M. Schopf and B. Nitzberg, Grids: Top Ten Questions., Scientific Programming, special issue on Grid Computing, 10 (2), August (2002), 103 - 111

[126] SETI@home, http://setiathome.ssl.berkeley.edu/ (01/03/2005)

[127] A. J. van der Steen and J. J. Dongarra, Overview of Supercomputers, 2004, http://top500.org/ORSC/2004/ (01/03/2005)

[128] Stone Souper Computer, http://stonesoup.esd.ornl.gov/ (01/03/2005)

[129] Booch, G., Object Solutions: Managing the Object-Oriented Project, Addison-Wesley, 1995

[130] Booch, G., Object-Oriented Analysis and Design with Applications, Addison-Wesley, 1993

[131] Lee, R. C. and Tepfenheart, W. M., UML and C++, A Practical Guide Object-Oriented Development, Second Edition, Prentice Hall, 2000

[132] Jacobson, I., Object-Oriented Software Engineering, Addison-Wesley Publishing Company, 1994.

[133] Quatrani, T., Visual Modeling with Rational Rose 2002 and UML, Pearson Education, 2002

[134] Rumbaugh, J., OMT Insights: Perspectives on Modeling from the Journal of Object-Oriented Programming, Cambridge University Press, 1996

[135] Rumbaugh, J., Blaha, M., Premerlani, W., Eddy, F., Lorensen, W., Object-Oriented Modeling and Design, Englewood Cliffs, N.J.: Prentice-Hall, 1991.

[136] UML, http://www.omg.org/uml/ (01/03/2005)

www.ingramcontent.com/pod-product-compliance
Lightning Source LLC
LaVergne TN
LVHW062307060326
832902LV00013B/2091